D0045530

SPLENDIDLY UNREASONABLE INVENTORS

SPLENDIDLY UNREASONABLE INVENTORS

The Lives, Loves and Deaths of 30 Pioneers Who Changed the World

JEREMY COLLER

with Christine Chamberlain

THE OVERLOOK PRESS
New York

This edition first published hardcover in the United States in 2009 by

The Overlook Press, Peter Mayer Publishers, Inc.
141 Wooster Street
New York, NY 10012

Copyright © 2009 by Jeremy Coller and Christine Chamberlain

All rights reserved. No part of this publication may be reproduced or
transmitted in any form or by any means, electronic or mechanical, including
photocopy, recording, or any information storage and retrieval system now
known or to be invented, without permission I in writing from the publisher,
except by a reviewer who wishes to quote brief passages in connection with
a review written for inclusion in a magazine, newspaper, or broadcast.

Cataloging-in-Publication Data is available from the Library of Congress

Manufactured in the United States of America
ISBN 978-1-59020-269-2
FIRST EDITION
10 9 8 7 6 5 4 3 2 1

Contents

Acknowledgements

I have much to thank Manchester University for, where I studied Management Sciences. One of my courses was on the history of the Industrial Revolution: it inspired me to ask, 'Who were the men behind these commercial inventions?' *Splendidly Unreasonable Inventors* has become a reality 30 years later. It was a collaborative process in the course of which a number of talented people contributed their expertise and advice. Richard Rivlin's selfless advice has been invaluable and as one of the first readers of early drafts he encouraged me to pursue my idea. Hugh Merrill has been enthusiastic and patient in keeping all of us on track and focused. Chris McDermott has, as usual, been a trusted and constructive critic. Christine Farrow-Noble served as a perceptive researcher. Charles Chamberlain spent long hours in his worldwide search for photography and drawings. Christine Chamberlain, who worked closely with me, was thoughtful and creative over the four years that it took to turn my original idea into a book. It has been a pleasure to work with her. Philip Blackwell encouraged and introduced me to my publisher, Infinite Ideas – the people there have been fantastic to work with.

My two children, Jodie and Jon, are inventing life anew and provide constant challenges and inspiration, and my wife Renée's encouragement and support have truly been the foundation on which any success I have had has been built.

Introduction

'There is no use trying,' said Alice.
'One can't believe in impossible things.'

When I say that I have produced a book on the lives of inventors, people ask why. My field is finance, miles away from the first telephone or the early sewing machine. Or is it? Invention and innovation occur in all disciplines and it is my thought that perhaps certain traits link people on the cutting edge, no matter what the product.

In my course on management sciences at Manchester University, one of our modules was industrial history and the commercial inventions that came to us from brilliant men like Arkwright and Stephenson, but something was missing. I was curious about the people themselves. I wanted to know what motivated them. The book I have written here is the book I wanted to read then.

The thirty inventors in this book are as different from one another as every human being is from his or her neighbour. Yet, for all their differences, they had something in common: an ability to 'see round the corner', and the motivation and drive to do something about it. Are there other traits that people with inventive personalities have in common? Are there life patterns that nurture and preserve the imagination and passion so often missing in day-to-day life?

A majority of these inventors and innovators lived unconventional lives and thought in uncommon ways. Because Jonas Salk, who developed the polio vaccine, was not accepted into the elite coterie of medical researchers, he was free to pursue his research in unconventional ways. Thomas Edison took nonconformity to the extreme, turning 'different' into an art form. None of the individuals you will read about here wasted time filtering the present through the past. They hungered for the future.

They were also on fire. Filled with a vision, they pursued their ideas with a single-mindedness that often bordered on obsession. Families were ignored, dinners lay uneaten, bills went unpaid. Some, like the Lumière Brothers and Marc Brunel, found a balance, but for most the combination of a compelling dream, the drive to get there first and the lack of adequate capital created tremendous pressure.

A brilliant idea fuelled by passion isn't enough. The hurdles to successful invention are many: shortage of funds, unlucky timing, vested interests, technical limitations, skullduggery, and – again and again – patent disputes. The system worked for some and against others. George Selden held up the entire automobile industry with a spurious patent, while Elisha Gray lost out to Alexander Graham Bell by a few hours, either through bad luck or illegal shenanigans in the Patent Office.

Neither do the fame and fortune of invention always attach themselves to the person who first lights on a revolutionary concept, or spends hours in the lab perfecting a process. Many of the people in this book were demonstrably not the first to come up with an idea, but they were the first to package the invention and it is their names that are written down in the history books.

Two things are necessary for inventors to make millions – their 'big idea' needs to work, and they themselves must be the conduit through which the world enjoys its benefits. In practice, many inventors were not, and are not, suited to bringing a big idea to market. They are too retiring, too idealistic about how things *ought* to be done or too perfectionist about the product of their labours. There are a number of examples in this book of very successful inventive partnerships, the prime example being the development of the Xerox process by Chester Carlson and Joe Wilson. In contrast William Shockley who worked at Bell Labs on the transistor, was consumed by the need to be the only one with his name on the patent and he ultimately stalled, while others around him, many working comfortably in partnerships, went on to develop Silicon Valley and make millions.

These stories are a canvas for understanding how human strengths and weaknesses determine outcome. All of us have dreams and ambitions. In looking at these extreme examples, you may find clues as to how others succeeded – or failed – in bringing their dreams to reality and the price they paid for success. Casting a glance backwards, perhaps we can learn more about ourselves, think more deeply about what we want from life and determine what price we are willing to pay to get there.

The sign above the doorway reads: ENTRANCE to READING and WRITING ROOM

That one big idea
KING CAMP GILLETTE

Born: January 5, 1855, Fond du Lac, Wisconsin
Died: July 9, 1932, Los Angeles, California

*Devise something that is useful and, once used,
requires the customer to come back for more.*

William Painter to King Gillette

AT THE END OF THE NINETEENTH CENTURY, King Camp Gillette was working for Crown Cork and Seal as a bottle-stopper salesman, but his heart wasn't in it. Instead, he was passionately committed to the concept of an 'earthly paradise', the Socialist utopia embodied in his 1894 book, *The Human Drift*. Gillette was also determined to make his fortune, an anomaly given his belief that economic competition, wealth and privilege were the source of all evil. It was later said that each side of King Gillette's personality 'seemed to have floated in separate, water-tight compartments'.

A natural tinkerer, Gillette held earlier patents, but, in his words, 'they made money for others, but seldom for myself'. By the 1890s, King Gillette was looking for a way to fame and fortune through one, important, invention. Harking back to his own, very successful, experience with the cork-lined bottle cap, Gillette's boss William Painter advised Gillette to devise something useful that, once used, was discarded, requiring the customer to come back for more. Painter was in a position to know: in one year the disposable bottle cap had earned him more than $350,000.

One summer morning in 1895, as forty-year-old Gillette was shaving with a razor so dull that it needed honing – requiring a trip to the barber or cutler – the idea came to him. Put the stropping and honing behind us, he thought, and replace this model with a disposable blade. The safety razor, employing thick blades made of forged steel that were resharpened when dull, had been on the market for about a decade. The heart of Gillette's new idea was a thin blade that was used two or three times, then replaced. What the salesman in Gillette really invented was the loss-leader, the idea of virtually giving away the razor to ensure long and continuous sales of the blades. And in the emerging razor market, it was up to Gillette to brand his product and establish himself as top of the line before anyone else got there.

'As I stood there with the razor in my hand,' he wrote later, 'my eyes resting on it as lightly as a bird settling down on its nest, the Gillette razor was born.' Without hesitation he sat down and wrote to his wife, "Our fortunes are made.'

His concept, an inexpensive, wafer-thin piece of sheet metal, sharpened on opposite edges, clamped down and set on a handle, didn't immediately instil enthusiasm. He was told by virtually everyone that his concept was a technical impossibility. 'How's the razor coming?' friends would ask sarcastically. King Gillette's only support came from his original mentor, William Painter who, after seeing a model, wrote to say, 'Whatever you do, don't let it get away from you.'

Over the next eight years, Gillette had to find an engineer to take on the seemingly impossible design project and backers willing to provide funds. MIT-trained chemist William Nickerson, a successful inventor in his own right, eventually designed the machinery that could mass-produce high-quality, low-cost blades. Three successful businessmen – Boston investor Henry Sachs, soft-drinks bottler Edward J. Stewart and shoe manufacturer and industrial promoter Jacob Heilborn – stepped forward to back the project. In the short term, this influx of capital was a great relief; in the long term, it marked the beginning of Gillette's loss of control over his patent and his future.

In 1901, the company was formed with King Gillette as president. He still owned the pending patent, but in 1902 he turned those rights over to the Gillette Safety Razor Company. Blocks of five hundred shares were offered at $250 and even at that price takers were hard to find. Those who did invest enjoyed dramatic returns. One fortunate buyer saw his $250 increase to $62,500 in four years.

Gillette may have been president of the new company that had formed around his invention, but he had essentially been sidelined by his investors. In order to earn a living, Gillette went to England as a salesman for Crown Cork, a dangerous remove from the heart of his company's activities. In the course of settling his family overseas, the original drawings and prototype for the first razor were lost. King Gillette was in England when he was alerted to plans to sell off overseas rights to the razor. Hurriedly, he boarded a ship and returned to Boston, where he succeeded in persuading the others to reverse their decision. Gillette, finally accorded a reasonable salary of $18,000 a year, resigned his job with Crown Cork. His immediate financial worries were over.

King Gillette tired of the company politics that swirled around him and, in 1910, accepted an offer of $900,000 for the majority of his shares in the company from brewery owner John Joyce. An additional incentive of $12,000 a year for five years plus the Gillette shares that he retained made King Camp Gillette a truly wealthy man for the first time in his life. Free at last of infighting and corporate responsibilities, King Gillette moved his wife and son Kingie to California, where he invested in real estate, including the eleven-hundred acre Gillette Ranch and a twenty-room, $125,000 mansion, which he gave to his son and daughter-in-law. He also bought nearly five hundred acres near Palm Springs and land parcels in downtown Los Angeles worth well over $2 million. The 'Razor King' ran three Pierce Arrow cars manned by two chauffeurs. Meanwhile, he travelled as the Gillette Company's goodwill ambassador.

In a stroke of marketing genius, each blade was packaged in a green wrapper featuring King Gillette's picture. As the most widely recognised person in the world, Gillette moved enthusiastically into his role as company ambassador. On a visit to Egypt, Gillette found himself surrounded by dozens of men, pointing at him while they simulated a shave. The Italian dictator Benito Mussolini was a dedicated Gillette user, as was Mahatma Gandhi. When pioneer aviator Major W.T. Black landed his plane in the Libyan desert in 1922, he found himself face to face with a nomadic tribesman wearing a Gillette razor suspended from his right car. It was no wonder that a Spanish gypsy displaying his collection of portraits of world leaders pointed to Gillette as the 'King of America'.

In 1917, for the first time in its history, the Gillette company sold more than a million razors. Blade sales soared to nearly ten million dozen. When King Gillette's initial seventeen-year patent expired on November 15, 1921, the company countered by introducing the 'New Improved' razor, turning what could have been a disaster into a commercial triumph.

While King Gillette rode high in California, far removed from actual company management, things were not going well in Boston. In early 1929, when King Gillette decided to sell stock in order to pay off massive real estate loans, he was dissuaded from doing so by the board of directors because they were worried the public would lose confidence in the company's stock.

It was a fatal decision. In 1929, a financial scandal erupted and the company that had once seemed indestructible, wavered. In October 1929, King Gillette, still holding all his shares, was hit by the market crash. His empire crumbled with frightening speed and one by one his properties went on the block for sums far below what he had paid. Gillette died, impoverished, in 1931.

Within thirty years, King Gillette had gone from rags to riches and back to rags thanks to one spectacular invention. He had persisted when people said it couldn't be done and his name and his face had become synonymous with the product. But King Gillette, with his utopian ideas, was no businessman and quickly lost control of his destiny. William Painter's advice – 'Don't let it get away from you' – had been far more critical than our inventor realised.

In 1932, a letter arrived at Gillette headquarters in Boston advising that the inventor's widow was living on a meagre $100 a month and requesting that the company, with earnings of over $100 million, contribute to her welfare. A company lawyer drafted a letter declining responsibility, but others thought better of the public relations impact and assigned $200 a month to Mrs Gillette.

Chester Carlson is pictured in the centre

One big chance
CHESTER CARLSON

Born: February 8, 1906, Seattle, Washington
Died: September 19, 1968, New York City

Invention was my one chance to start with nothing and end up with a fortune.

Chester Carlson

THIS IS THE STORY OF AN INVENTION that preceded a perceived need – that is to say that when consultants in the 1950s failed to detect a market for the nascent Xerox office copier, the mightiest of corporations listened. Time and again the likes of IBM and 3M turned down the chance to take part in the Xerox project. Things were fine just as they were, thank you very much. Office workers were already basking in the accelerated capacity afforded by the A.B. Dick mimeograph. What more did they need?

The fact that there was as yet no language to discuss the making of copies at the push of a button was a major impediment. The consultants got the wrong answers because they asked the wrong questions. Would you spend more than $2000 for a copier? In the abstract, that seemed like a lot of money and the answer was no; in practice the convenience factor would far outweigh the cost.

Fortuitously, Chester Carlson and a few others, including Joe Wilson whose company Haloid developed the copier, chose to ignore consultants' advice. Billions of dollars in profit later, their decision was validated.

Chester Carlson's inventiveness and tenacity took root in a childhood defined by poverty and isolation. In 1910, four-year-old Chester, an only child, was living with his family in a tent in the sweltering Arizona desert. His father, an itinerant barber, invested what little savings they had in a land scheme in Mexico, and the family moved south to a two-room adobe hut on a thankless piece of ground. Chester was one of two kindergarten students in a makeshift school.

Abandoning that forlorn home in the face of the Mexican Revolution in 1911, the Carlsons sailed up the coast to Los Angeles. Destitute, the family settled into a small room behind the house where Ellen Carlson worked as a housekeeper. Moving again to a virtual shed in the mountains near San Bernadino, Chester registered as the only student at the one-room schoolhouse. During recess, he wandered alone in the dusty yard. By the time he was eight, he was working at whatever menial job he could find. By the time Carlson entered high school, he was the family's principal provider.

Carlson acknowledged later that the solitude of his childhood provided a lot of time for creative thought and rendered him self-reliant and comfortable with his own company. The long years that he shouldered responsibility for his family, waiting for his chance in life, instilled patience. All of those qualities came into play later.

At the age of seventeen, when his mother died, Chester and his father moved into an abandoned chicken coop. Chester had given up on the idea of school until an uncle advised him to continue his education at any cost and encouraged him to take advantage of an affordable four-year work-study programme at Riverside Junior College in California.

Juggling time and funds, Chester managed to graduate from CalTech in 1930. His first job, in the patent department at Bell Labs and, later, his work for a patent law firm in New York, not only armed him with the tools to write strong patents later for his own inventions, but also reinforced his conviction that there was a very real need for new methods of making copies in busy offices. His eye was on an entirely new kind of copying process, one in which light was the key element. Carlson had profound faith, both in the potential market and in his idea, and he pursued that conviction to fruition over the next twenty years.

By 1937, Carlson was envisioning a process that combined electrostatics with photoconductivity, using an electrostatic charge on a surface dusted with fine powder to create an observable image. By his own admission, he had a 'big idea by the tail', but he didn't know what to do with it. In 1938, with his assistant Otto Kornei, Carlson produced the first xerographic copy. It didn't amount to much and Kornei, unimpressed, moved on.

Carlson put together a rudimentary demonstration kit, but IBM, 3M and A.B.

These words, written on a microscope slide, represent the first 'photocopy'.

Dick were among those who exhibited what Carlson referred to as an 'enthusiastic lack of interest'. IBM's rejection of Carlson's offer of an exclusive licence in exchange for a five per cent royalty represented the first of many opportunities that the corporate giant had to capitalise on what *Fortune* magazine later described as 'the most successful product ever marketed in America'.

Nineteen-forty was a bumper year. Carlson was issued the first of four basic patents, a patent so all-encompassing that it was later described as 'one of the best US patents ever written'. Graduating from law school, he soldiered on alone over the next four years to market electrophotography, with no success. Carlson's luck changed in 1945 when the head of the graphics art division at the non-profit research organisation Battelle Institute described Carlson's idea as a 'good research gamble'. Battelle became the exclusive agent for Carlson's patents in exchange for fifty per cent of future royalties. Eight years had passed since the first basic copy had been produced.

In the search for a production partner, giants like Kodak continued to be dismissive. Then, one hot summer's day in 1945, a Haloid Company executive was reading in the men's room and happened on an article in the Kodak *Bulletin* on electrophotography. He was sufficiently interested to bring the article to the attention of his boss, Joe Wilson. Haloid, a small, family-owned business in Rochester, New York, producing photographic and copy paper, was looking for something new. Encouraged by their employee's recommendation, they sent two engineers to evaluate Carlson's idea. The engineers reported back that the process had 'tremendous possibilities' and 'should not be overlooked'. To give some idea of the crude nature of that early model, the 'machine' involved a transparent plastic ruler, a metal plate, a bright light and a piece of cat's fur.

Joe Wilson and Chester Carlson met for the first time in 1946. Wilson, a visionary leader, a gentleman and a man of ideas who read voraciously, instantly appreciated Carlson, a quiet but tenacious scientist. They shared an understanding that this invention was the one big chance for both of them. There was no corporate swagger here, just good instinct and common sense. The development partnership lasted more than twelve years and was to nearly bankrupt Haloid.

In the mid-1950s Haloid began work with Carlson on developing the 914 model. Carlson continued to experiment in his basement, transporting soda pop bottles filled with toner down the street in the early morning. During this time, Carlson added thirty-six additional xerography-related patents to his four original applications. Physicists and scientists who worked with Carlson found him to be unassuming, generous about sharing credit, humble about his accomplishments, brilliantly inventive and doggedly persistent. Given his isolated upbringing, Chester Carlson was better than most inventors at working with others.

In 1954, IBM came nosing around, looking for an exclusive licence to manufacture the xerographic office copier. Fortunately, IBM asked for more than Haloid was willing to give away. IBM backed off and Haloid continued on its own. At one point, IBM hired influential consultants Arthur D. Little who reported back that a copier like Haloid's 914 'had no future in office copying markets' and IBM backed out for good. IBM's Tom Watson said later that his failure to buy out Haloid in the 1950s had been his biggest mistake.

Haloid, which changed its name to the catchier 'Xerox' in 1958, hired their own consultants, Ernst & Ernst, who returned the pessimistic view in 1959 that there was no future for the copier. With the support of his board, Wilson went ahead with the 914. To go against the advice of two of the largest international consultancy firms and ignore reservations expressed by corporate giants in the photographic and copying industry represented a great leap of faith. It may also have been a case of being in too far financially to turn back. 'This just goes to show,' wrote Russell Dayton of the Battelle Institute, 'if you've got something unique, you don't take a poll.' A successful Xerox machine, like many inventions before and after it, would both create the need and fill it.

Joe Wilson was his own man and a broad thinker and the decision to move forward was ultimately his, but he was certainly influenced by the quiet confidence and self-reliance of his partner, Chester Carlson. Wilson was justifiably anxious about his financial position and Carlson was frustrated by the slow progress, but these tensions were defused by a mutual courtesy. The team of Haloid engineers were tired and frustrated, but they were also stimulated by the project and determined to find solutions. People arrived at the factory at all hours of the day and night to try out a new idea.

By the end of the 1950s, Haloid had built fifty 'reasonably functional' 914s and the factory began to turn out machines at the rate of five a day. Over fourteen years, Haloid had invested more than $12 million in the project, a sum that exceeded the company's entire earnings in the 1950s. The 914, capable of turning out 100,000 copies a month without the use of chemicals, was introduced in 1960.

When outside consultants prepared their reports on the long-term viability of the Xerox machine, one of the critical factors was the high price tag on the new machine. In a magnificent marketing coup, Haloid decided that paying per copy would be a more palatable alternative and installed meters. In 1962, Haloid shipped ten thousand copiers.

Chester Carlson saw invention as his one big chance to be truly successful financially, but when he found himself with a fortune exceeding $150 million in the 1960s, life changed very little. He and his wife remained in their modest home. There were no country club memberships, no second car, no yachts, no airplanes. He remained unassuming in manner. On one occasion when Chester mentioned that he worked for Xerox, the listener took for granted that he was a factory worker and asked Chester if he belonged to the union. As money poured in, Carlson admitted to feeling lost and purposeless. Eventually, he focused his attentions on the thoughtful distribution of his millions. Most of that giving, amounting to over $100 million, was anonymous. Deeply spiritual and reflecting on his life, Carlson wrote, 'I do see that some of the sufferings I went through, and that my parents went through, were blessings in disguise which have enriched and deepened my life and have finally provided a philosophy which is a raincoat in time of difficulty.'

'His real wealth,' his wife said later, 'seemed to be composed of the number of things he could easily do without.'

In New York for a meeting, Chester found himself with a few free hours and bought a ticket for the matinée of the British film, *He Who Rides a Tiger*, starring Judi

Dench. During the performance, in the dusky theatre, Chester Carlson died of a heart attack. He was sixty-two. The day before he died, Chester had bought a balloon just outside Central Park and, letting it go, watched as it floated over New York. 'Chester tiptoed into life,' his wife said later, 'and tiptoed out again.'

Speaking of his friend and partner in the great Xerox saga, Joe Wilson quoted George Bernard Shaw: 'All progress depends on the unreasonable man, because reasonable men accept the world as it is, while unreasonable men persist in adapting the world to them. Chester Carlson,' he added, 'was splendidly unreasonable.'

Three years later, Joe Wilson was attending a luncheon given by Nelson and Happy Rockefeller at their Fifth Avenue apartment. When Happy looked to her left, she noticed that Wilson had slumped down in his seat. Dead from a heart attack at sixty-one, Joe Wilson never regained consciousness.

A toad in his time

ISAAC MERRITT SINGER

Born: October 26, 1811, New York City
Died: July 23, 1875, Devon, England

The most important thing for an innovator
isn't necessarily being first. It's being able to put
together a combination that works.

Harold Evans, *They Made America*

'I DON'T GIVE A DAMN FOR THE INVENTION, the dimes are what I'm after,' said Isaac Merritt Singer. In the end, Singer had plenty of dimes. At the age of forty, he was virtually penniless; by the age of forty-four he was worth millions, thanks to the sewing machine.

Life on the stage was Isaac's real dream. Biding time until he could pursue a serious acting career, Singer worked for a while as a labourer on the Illinois & Michigan Canal, where he invented a machine for drilling rock that earned him $2000. Money in hand, Singer returned to the theatre with his troupe, the Merritt Players. He appeared on stage as Isaac Merritt; his girlfriend, Mary Ann Sponsler, appeared as 'Mrs. Merritt'.

Meanwhile, another stage had been set on the industrial front. In addition to earlier prototypes produced in France, England and Germany, Americans Walter Hunt (who also invented the safety pin) and Elias Howe had designed a sewing machine that incorporated three vital components: the lockstitch (created by using two separate thread sources), the eye-pointed needle and a shuttle.

Enter stage left thirty-eight-year-old Isaac Singer, who wandered into a New York workshop in June 1850 to find machinist Orson Phelps struggling to construct a sewing machine. Perplexed by design flaws, Phelps was close to giving up. At the time of this meeting, things were not going well for Singer either, who was living hand to mouth in two rooms with his common-law wife, Mary Sponsler, and their children. Singer's invention of a machine for carving printer's type out of metal and wood, had come to an unfortunate end when an exploding steam boiler destroyed the only prototype and sixty people with it. A second prototype had attracted little attention.

Phelps' suggestion that Singer would be better served by perfecting the sewing machine was met with little enthusiasm. 'What a devilish machine,' exclaimed Singer, according to Phelps. 'You want to do away with the only thing that keeps women quiet, their sewing.'

In spite of his reservations, Singer quickly identified the design flaws and sketched a new model. Three men – Phelps who was to contribute his mechanical skill, George Zieber whose job it was to attend to the business side of things, and Singer, who would contribute his inventive genius – became co-partners in I.M Singer and Company.

Phelps put the machine together, but it still didn't work as it should. On their way home late that night, Zieber mentioned the loose loops of thread on top of the material and Singer, in a flash, understood that they need only adjust the tension of the needle thread to perfect the product. Returning to the workshop at 1:00 a.m., they worked through the night. By 3:00 p.m. the problem had been solved.

Singer's machine, first patented in 1851, was superior to all previous models, with a needle that moved up and down rather than sideways. Later Singer devised a treadle that freed the operator's hands and worked out the means for sewing a curved as well as a straight line. Manufacturing began in late 1850, not long after Singer had first seen the sewing machine in Phelps' workshop. George Zieber suggested changing the machine's name from the Jenny Lind to Singer, taking advantage of the implication that this new invention literally sang.

Singer, who saw himself as the true father of the sewing machine, applied for a patent in his name only, thus gaining full control of the invention. With money gathered from the initial sale of thirty machines at $100 each, Singer bought out Orson Phelps for $4000. When Zieber, a generous, good-hearted, and somewhat naïve individual, enquired what part of Phelps' interest would accrue to him as the

A mid-nineteenth-century version of Singer's sewing machine.

third partner, he was informed that he had been given more than enough already.

Initial sales were slow. Selling the idea of the sewing machine wasn't necessarily a straightforward task. In 1830, when a functional sewing machine was introduced by French tailor Barthelemy Thimonnier, tailors, fearing for their livelihood, stormed and destroyed Thimonnier's eighty-machine plant. The inventor fled and later died a bankrupt.

Singer, a dramatic character and a born salesman, began to circulate around country fairs, where he engaged pretty young women to demonstrate the sewing machine. Singer's sense of theatre served him well when he gave a grand Invitational Ball for three thousand guests, during which five pretty women stitched the evening away. In order to emphasise that sewing machines were suitable for respectable women, Singer sold machines at half price for use in church sewing circles. The cabinets, some in the shapes of animals or cherubs, were designed to emulate high-quality pieces of furniture.

Among the people who caught a Singer display was a despondent Elias Howe, who held the patent for the all-important lockstitch incorporated by Singer. Redeeming his patent from a London pawnshop, Howe demanded a payment of $25,000. Singer, who didn't have the money, gave way to his towering temper and threatened Howe.

Not surprisingly, Howe sued and Singer was forced to sign over a third of his business – the third he had bought from Phelps – to pay for legal representation. Singer's lawyer Edwin Clark was cold, calculating and socially respectable, and his association with Singer proved to be ruthlessly creative. Clark and Singer agreed to split the rights to Patent No. 8294 for the Singer Sewing Machine 50:50, excluding the third partner, George Zieber. In exchange, they promised Zieber a third of the profits.

In December 1851, confined to bed with a high fever, Zieber was advised by Singer that he would not recover and agreed to sign over his rights to the sewing machine for the sum of $6000. It must have been a bitter pill when the company assets grew to $500,000 within the first years and Zieber found himself doing small jobs for Singer in order to earn extra money.

Singer dismissed Elias Howe as a 'humbug', but Howe did not give up easily. After a bitter engagement described in the papers as the Sewing Machine War, Singer and Clark were forced to pay Howe $25 on every machine sold. Hundreds of new sewing machine patents resulted in endless and costly litigation, a problem that was solved when, in 1856, seven manufacturers and Elias Howe agreed to pool their patents to form the Great Sewing Machine Combination. The Combination received $15 for every sewing machine sold and Elias Howe, who received $5 of that, became a rich man.

Singer, aided by Clark's superlative marketing sense, did well in the competitive market. Lightweight machines for home use, the introduction of long-term instalment purchase plans, trade-in options, elaborate showrooms with instructors and salespeople, foreign-based manufacturing, and efficient distribution of parts gave Singer machines a monopoly on three-fourths of the entire world market.

Singer's personal life was less smooth. A colourful and dramatic figure, Isaac Singer had at least six wives, common law and legal, three of whom were named Mary.

He established families under at least three different names – Matthews, Merritt and Singer – and his brood included twenty-four or more children born out of wedlock.

In 1860, Singer's past caught up with him. Riding down Fifth Avenue in her carriage, Mary Ann Sponsler – a faithful consort who had lived with Singer for more than twenty years and borne him numerous children – noticed a happy couple riding in the opposite direction. That happy couple, Mr and Mrs Matthews, was, in fact, Singer and his common-law wife Mary McGonigal, with whom he had five children. Mary Ann Sponsler spared no words as she told Singer what she thought. When Mary Ann returned to her palace at 14 Fifth Avenue, she found an enraged Singer who, in the throes of his terrible temper, beat her senseless. Their daughter tried to intervene and was rendered unconscious by her father.

Convicted of brutality, Singer escaped to Europe, where he met his fifth and final wife, Isabella Boyer, then in her twenties to Singer's fifty-one years. They wed in 1863 and had six children.

Singer's attorney Edward Clark, for whom social standing was of paramount importance, was appalled by the scandal and moved quickly to disassociate himself from Singer. When the partnership was dissolved, Clark gained control of twenty-two patents and over $500,000 in assets. Singer, with 40% of the stock, was left with an assured income.

Back in New York, Singer and his wife Isabella were easy to pick out as they rode through Central Park in their outsized canary-yellow coach drawn by nine horses and driven by liveried footmen, but money wasn't enough to gain acceptance. Snubbed by New York society, they retired first to Paris and then, escaping the Franco-Prussian War, to Paignton, on England's Devon coast, where Singer remodelled Oldway Mansion in the style of a French villa. He called his new palace 'Wigwam'. When Singer died on July 23, 1875, two thousand people lined the roads, church bells tolled and flags flew at half-mast. His estate totalled $14 million.

Singer's eighteenth child, Winnaretta, a prominent patron of avant-garde music, married Prince Edmond de Polignac in 1893; it is said that she modelled for the Statue of Liberty. Another daughter, Isabelle-Blanche, married the duc Decazes et Glücksberg, and had a daughter, Daisy Fellowes, a London socialite and one-time editor of *Harper's Bazaar*. Isabelle committed suicide in 1896.

One of Isaac Singer's sons, Paris, had a child by Isadora Duncan, the famous actress and dancer who choked to death when her scarf caught in the rear wheels of her car. Paris remodelled the Wigwam in the style of Versailles after his father died, only to lose most of his remaining fortune in the Crash of 1929. Another son, Washington, contributed handsomely to what is now the University of Exeter, where a building is named in his honour.

Before we leave Isaac Singer and his sewing machine, we should look at two other players in this saga, Walter Hunt and Elias Howe. Both came very close to success with the sewing machine, but they lacked two things that Singer brought to the table: a flair for marketing and ruthless ambition.

Walter Hunt (1796–1859) floated in a world of pure invention. His chronic indebtedness and lack of business sense led directly to poor decisions, a good example of which was the sale of his 1849 patent for the safety pin for $400. More importantly for our story, Walter Hunt, working in his small shop in New York City, put together a sewing machine that incorporated two key features of Singer's later model. He shelved his invention when his fifteen-year-old daughter persuaded him that the sewing machine would put too many people out of work. At the time, more than ten thousand women in New York alone earned their living working in the factories.

Embroiled in an ugly court battle with Elias Howe over the sewing machine patent, a fight that lasted from 1849 to 1854, Singer and his lawyer offered the impoverished Walter Hunt $50,000 to recreate his original sewing machine. If they could prove that Howe had *not* been the first to perfect the machine and the lockstitch, his patent would be null and void and their case would be won. Try as he would, Hunt could not reproduce his original invention and the reward slipped tantalisingly through his fingers.

Walter Hunt was ever the inventor, and over the course of more than thirty years was awarded twenty-six patents, including paper collars, a streetcar bell and a knife sharpener. Hunt's Antipodean Apparatus, a pair of special shoes, allowed a performer to walk on the ceiling, but only after some training. 'Hunt's Restorative Cordial' – 'The Life Preserver' – was a popular tonic of the day. Unfortunately, none of these inventions made Hunt a wealthy man. He died in 1859, leaving his children in a state of genteel poverty. In one obituary, Hunt was referred to as 'the Bonaparte of invention', with the caveat that the sewing machine was his 'Austerlitz', a reference to Napoleon Bonaparte's greatest military victory.

Elias Howe concentrated his energies on designing a sewing machine, but had neither the capital for manufacture nor the savvy to overcome the lobby against mechanisation. He gave up on his invention, but not before he had patented the lockstitch and the eye-pointed needle. When Howe failed to market his sewing machine in America, he took his family to England, where he sold the British rights to the sewing machine to a corset-maker named William Thomas. Thomas took out a British patent and garnered a fortune, leaving Howe with nothing. By 1849, a penniless Howe pawned his US patent papers and returned to the United States.

Back in America, Howe discovered that Singer's machines, based on his fundamental patents, were a wild success. Howe sued and won. Howe was awarded $2 million, the wealth he had dreamed of years before, but the struggle had taken its toll. In 1867, the year his patent expired, Howe died. He was forty-eight years old.

Elias Howe also obtained a patent in 1851 for an 'automatic, continuous clothing closure.' Perhaps because he was preoccupied with his lawsuit involving the sewing machine, Howe missed the chance to make his name as 'father of the zip.'

Pearls of wisdom
KŌKICHI MIKIMOTO

Born: March 10, 1858, Toba City, Mie Prefecture, Japan
Died: September 21, 1954

There are two things that can't be made at my laboratory:
diamonds and pearls.

Thomas Edison

THE EXPECTATIONS WERE THAT KOKICHI MIKIMOTO, the eldest of five children, would one day take over his father's noodle shop, passed down through the generations. In anticipation of his role as breadwinner, Mikimoto left school at thirteen to sell vegetables, but his fascination with pearls grew as he watched the famous divers of Ise unloading their harvest. At a young age, he was aware that from the thousands of oysters harvested it was rare to find even one large pearl. He also understood that demand was outstripping supply.

At the age of thirty, Mikimoto obtained a loan and, together with his wife Ume, set out to do what no one else had done: entice oysters to produce round pearls on demand. As early as the eighth century, the Chinese had introduced small lead Buddhas into oyster shells to trigger the flow of nacre, producing beautiful pearl-covered figures, but the art of cultivating the highly coveted round pearl remained elusive. Gathering together a team, Mikimoto established a pearl farm in the Japanese coastal areas of Ago and Toba, where he carried out his first experiments. When early attempts failed, the team disbanded in frustration. Only Ume remained at his side.

Nature seemed to work against Mikimoto in myriad ways but, after years of experimentation and near bankruptcy, he met with success. In 1905, in the wake of a red tide that destroyed eight hundred and fifty thousand oysters, Kokichi Mikimoto opened a shell to find, at last, a perfectly round pearl, as perfect as any natural pearl he had seen.

Ready to do business, Mikimoto now discovered that he was not the only player in the field. Tokishi Nishikawa, a government biologist, in partnership with a carpenter, Tatsuhei Mise had spent time in Australia, where they learned the secret of spherical pearl production from a British marine biologist. The process of culturing a pearl requires the insertion of a piece of oyster epithelial membrane, together with a nucleus of shell or metal, into the oyster's body, causing the tissue to form a pearl sac. That sac, developed in response to the irritant, secretes layers of nacre, a pale substance composed of microscopic calcium carbonate crystals, that coats the nucleus – thus creating a pearl. In 1907, Mise was awarded a patent for a grafting needle; when Nishikawa applied for a patent for nucleating, he discovered that, working independently, he and Mise had discovered the same principle. In a compromise unusual in the world of invention, they joined forces to patent the 'Mise–Nishikawa Method', which remains the heart of the cultured-pearl process.

Mikimoto held three patents related to the technique for culturing round pearls in mantle tissue, but none of them was commercially viable. It was only when Mikimoto bought the rights to the Mise-Nishikawa Method in 1916 that he was able to eclipse all other participants in the field. The culturing of pearls became a family business when Tokishi Nishikawa became Mikimoto's son-in-law.

The process of inserting the nucleus into the oyster isn't an easy one and a fairly high percentage of treated oysters die. Mikimoto experimented with various substances, including glass and wood, but found that round nuclei cut from American mussel shells found in the Tennessee and Mississippi Rivers were the most effective irritant. His process provided the basic principle for the farming of virtually all salt-water cultured pearls in the twentieth century. To illustrate the complexity of the process, out of a bed of three million oysters, about 5% yielded a pearl after seven to nine years. Mikimoto was known to burn shovelfuls of imperfect pearls for an audience of journalists.

Kokichi Mikimoto attracted a lot of attention when he opened the doors of his second retail store in Tokyo's Ginza district in 1906. In this two-storey, Western-style building made of white stone, stylish young men dressed in finely tailored, high-collared three-piece suits waited on customers. New displays, redesigned monthly to the highest standard, reflected Mikimoto's appreciation for Western aesthetics combined with his own, unique, sense of style. So successful was the venture that he soon opened stores in London (1913), followed by Paris, New York, Shanghai and Bombay. Scouts were dispatched to the fashion centres of the world to research trends and styles. 'I would like to adorn the necks of all the women of the world with pearls,' was Mikimoto's modest ambition.

In 1926, Mikimoto exhibited his pearl-covered Liberty Bell at the World Exposition in Philadelphia. When he met inventor Thomas Edison, an avid pearl collector, the following year, Edison told Mikimoto that only two things could not be made at his laboratory – diamonds and the pearls he so loved.

During the Second World War, B-29s levelled Mikimoto's fabulous Tokyo showroom and strafed his Ago Bay factory, but he still had half a million oysters in the bay and a fortune in pearls in boxes scattered around his home. He didn't hesitate to raid his hoard to sell to the eager GIs. 'I like Americans best,' he was heard to say. 'They are so straightforward – like children.'

Kokichi Mikimoto, the 'Pearl King', spent his last years in a modest, unpainted four-room house overlooking Ago Bay on Mikimoto Pearl Island. If visitors expected to find a regal figure, what they discovered instead was a small, wizened man, wearing a black bowler hat and a brown kimono. In 1946, eight years before his death and in the first year of the American occupation, Mikimoto declared the largest personal income in Japan: he had netted over 3 million yen selling pearls to the conquerors.

The gift of addiction
JOHN STITH PEMBERTON

Born: January 8, 1831, Knoxville, Georgia
Died: August 16, 1888, Atlanta, Georgia

Don't touch this business.
Not one man in a thousand ever succeeds in it.

Edison, to an aspiring inventor

PATENT MEDICINES WERE A FORM OF INVENTION in the nineteenth century that brought fame to many, including John Pemberton, who gave us Coca-Cola. A college graduate with a degree in pharmacy, Pemberton focused his attention not only on patent medicines, but also on the manufacture of hair dyes, cosmetics and perfumes. A respected citizen and pharmacist, he purchased interests in several chemists shops and established the J.S. Pemberton Company.

In his early fifties, Pemberton turned his attention to the coca plant. Touted as a stimulant, digestive aid, life-extender and aphrodisiac, coca had been chewed by the natives in Peru for over two thousand years. In addition to vitamins, proteins, calcium, iron and fibre, the coca leaf contains a natural pesticide, cocaine, which inhibits the reuptake of dopamine in humans and leads to addiction. Pemberton's improvement on existing coca tonics involved the addition of the kola nut, also a strong aphrodisiac, with potent medicinal properties. In addition to a longing for fame and fortune, Pemberton was hopeful that this new elixir might cure his long-standing morphine addiction, a legacy of his service in America's Civil War.

Pemberton mixed the first batch of coca-kola syrup in a brass kettle in his back yard. Visiting John Pemberton's house in Atlanta in the 1880s must have been quite an experience An enormous filter extended through the top two floors of Pemberton's house and the back yard was filled with vats for mixing and boiling. The back-room laboratory, where the secret ingredient was tested, was off limits to visitors. Pemberton's nephew characterised his uncle as 'secretive' and 'obsessed', often missing meals and working late into the night.

'Pemberton's French Wine Cola', launched as an 'intellectual drink' and an 'invigorator of the brain' in 1885, was initially marketed by local chemist shops as a tonic. When the kola nut was added, the new brew was advertised as a 'nerve tonic' capable of curing morphine addiction, a common complaint among both doctors and patients. When Atlanta banned the sale of alcohol, Pemberton made some creative changes to his formula and dropped the reference to wine, billing his new concoction as a 'temperance drink'. He earned more per day with his new brew than he had previously earned in a year.

The thick brown syrup proved to be a greater obsession for John Pemberton than his morphine addiction. As the drive to create the perfect drink consumed him, Pemberton shut himself away in the laboratory, forgetting to eat and paying little attention to his family and business affairs. The magic cure for morphine addiction didn't work for Pemberton, who slipped further into ill health. As a result of endlessly testing his concoction, he was now addicted to cocaine as well as morphine. He also suffered from a gastro-intestinal complaint that might well have been cancer.

As Pemberton created and recreated his syrup, he sent samples to shop-owner Willis Venable, who mixed the brew with soda and tried it out on customers. The first true Coca-Cola mix was served on May 8, 1886 for 5 cents a glass. The first advert for Coca-Cola appeared in the *Atlanta Journal* on May 29. In 1887 Pemberton registered the Coca-Cola trademark in his name. The syrup and the name were never patented, illustrating an instance in which the protection offered by a trademark proved more effective than a patent. In the course of applying for a patent, the magic recipe would have been revealed and laid open to competition when the patent ran its course. Coca-Cola remains one of the best-kept trade secrets in history.

Pemberton was no businessman. Believing erroneously that he would never again rise from his sickbed, he further compromised any hopes of financial success by selling two-thirds of his company to Willis Venable and George Lowndes. His claim that Coca-Cola would one day be a national drink indicates that Pemberton understood the potential of his creation, but he was ill-equipped to protect his interests. Pemberton did, however, retain a third of the shares to provide a living for his son.

In 1888, Atlanta pharmacist Asa Griggs Candler purchased all rights to the formula from Pemberton and his two original partners. In constant need of money to finance his addictions, Pemberton double-sold shares totalling $2000 to a group of three investors. That second tier of investors lost their money. By 1891, Candler had wrested control of all Coca-Cola shares for a total investment of $2300. Later handwriting analysis indicated that Pemberton's signature on Candler's bill of sale was most likely a forgery.

Candler replaced the cocaine with other, more 'wholesome' ingredients and by 1895, less than ten years after Coca-Cola was first served in Willis Venable's corner shop, there were Coca-Cola factories in Dallas, Chicago and Los Angeles. By 1900, Asa Candler was one of the richest men in Atlanta.

In 1888, at fifty-seven, John Pemberton died, leaving very little money in his estate. His widow, Ann, died in poverty. Pemberton's son, who at one time was in line for one third of the company, died in 1894 at forty, either by his own hand or as the result of a morphine overdose. On the day of Pemberton's funeral, Atlanta pharmacists paid their respects by closing their shops and attending the service en masse. On that day, not one drop of Coca-Cola was dispensed in the city.

Mr Dynamite
ALFRED BERNHARD NOBEL

Born: October 21, 1833, Stockholm, Sweden
Died: December 10, 1896, San Remo, Italy

I hold my inventive faculty on the stern condition that it must master my whole life, often have complete possession of me, make its own demands upon me, and sometimes for months together put everything else away from me.

Charles Dickens

ONE APRIL MORNING IN 1888, Alfred Nobel opened his newspaper and was stunned to read his own obituary. Perhaps even more shocking was to see himself described as the 'merchant of death'. With this unusual insight into posterity, Nobel had a chance to set things right: eight years later, in 1896, the man who had spent his life trading in explosives and armaments would designate the bulk of his fortune to the recognition of achievements in the sciences and literature and the pursuit of world peace.

Alfred's father, Immanuel, an uneducated man whose fortunes swung alarmingly between bankruptcy and affluence, invented the process for making plywood, designed a boiler-driven hot water heating system, and made drawings for a coffin complete with a cord attached to a bell and air holes in case of premature burial. In so doing, he touched on his son Alfred's worst fear, that of being buried alive.

What Immanuel did not have in large measure was business sense. Bankrupt, he fled debtors' prison in Sweden and crossed the sea to Russia in search of opportunity. Alfred's mother, Andriette, was left to fend for herself and her three sons. Alfred, aged four, a frail and chronically ill boy, was reduced to selling matches on street corners with his brother, Ludvig. The boys attended a school for the poor, a cold and forbidding institution that believed in corporal punishment.

When the family finally joined Immanuel in St Petersburg five years later, they found him thriving on profits derived from land and underwater mines and the proceeds from a large factory and foundry. Alfred and his brothers studied under private tutors. By the mid-1850s, with the Russian defeat in the Crimean War, life changed again and the Nobels went bankrupt for a second time. The sense of humiliation stayed with Alfred Nobel all his life, contributing to his isolation, melancholy and sensitivity to ridicule.

Through his tutors, Alfred was aware of the Italian inventor Ascanio Sobrero's work with nitroglycerine, an oily, colourless unstable liquid used in the manufacture of explosives. Sobrero, whose face had been badly scarred in a laboratory explosion, had determined that this powerful explosive was too dangerous to be of any practical use and abandoned his experiments. Nobel took up Sobrero's work, applying creativity and imagination to this volatile mix. The first step was the addition of gunpowder to the nitroglycerine, creating a far more powerful explosive than either substance could claim to be on its own. In 1863, a patent was issued to Nobel for 'gunpowder for both blasting and shooting'. With canals and bridges, mining interests and railway systems sprouting up around the world, these explosives had great industrial potential and Alfred, who was determined to turn his inventiveness into a profit-making venture, saw the possibilities. One day, while Alfred was away from the Stockholm lab, an explosion killed his youngest brother and three assistants. Alfred continued his work the next day in spite of the personal loss and the public outcry. That explosion was known long afterwards as the Nobel Bang. Immanuel Nobel, grieving for his dead son, suffered a debilitating stroke a few weeks later.

Alfred had been right, there was high demand for the new and powerful explosives, but there was a price to pay. Nitroglycerine, transported haphazardly in tin cans, was an accident waiting to happen. An officer's servant unwittingly used the explosive to oil boots and harness. Nobel himself usually travelled with a few cans in his luggage and Orlando Webb, chairman of the board of a Welsh mining company, defied the British prohibition on the transport of explosives by storing nitroglycerine in wine bottles packed in crates marked 'light wine from the Rhône country'.

When a German traveller left ten pounds of explosives at his New York hotel, the case was used as a seat by the porter until, one day, he noticed smoke and threw the bottles into the road. The ensuing eruption destroyed neighbourhood buildings and injured residents and pedestrians. An explosion on the steamer *European* killed nearly fifty people, and Nobel found himself vilified in the newspapers as 'the salesman of death' and a mass murderer.

Step by step, Alfred Nobel addressed the safety issues, notably with the invention of the blasting cap or detonator in 1863. During the 1860s, he also patented 'Nobel's Safety Powder', nitroglycerine combined with a porous silicate to render it more stable. When he cut his finger in his Paris laboratory, Nobel applied collodion, a highly flammable, syrupy solution of cellulose nitrate, ether and alcohol. Awakened at 4:00 a.m. by the pain in his hand, he went to his lab and, with the collodion in mind, began to consider the possibilities for an explosive that would be even safer than dynamite. The result, blasting gelatin, a mixture of nitroglycerine and nitrocellulose, produced an explosive that was insensitive to shock, safe when lighted and impervious to water.

It was conceivable to justify the development of explosives as beneficial to humankind. Less defensible were the armaments, including ballistite, a smokeless gunpowder, that Nobel patented and produced in the 1880s. Nobel believed that potent weapons were actually a deterrent to war, but when terrorists used dynamite to kill Tsar Alexander II of Russia and soldiers carried new and more deadly weapons into the Franco-Prussian War, it became increasingly difficult to reconcile pacifism with invention. Highly intelligent, well read and rarely distracted from his work by social pursuits, Alfred Nobel had a prodigious talent for finance and business. Wealth for its own sake did not interest him; the determination to succeed and a highly developed sense of responsibility to those with whom he was involved were the greater motivation.

The explosives Nobel patented became the foundation of a global empire. In partnership with his two brothers, Alfred was also a partner in the Nobel Brothers Naptha Company, which developed the rich Baku Oil Fields on the Caspian Sea. By the mid-1880s, the Nobels had thirty-two large tankers, fifteen-hundred tank trucks, and employed more than five thousand people. That empire would endure until the Russian Revolution in 1917 put an end to exploration.

Short in stature, serious of expression, bearded and habitually dressed in a dark suit, Alfred Nobel's personal life did not reflect the decisiveness and mastery he displayed in business matters. Fluent in five languages, Alfred Nobel disciplined himself by translating Voltaire from the French into Swedish and back again, comparing his final translation with the original text. The owner of five houses and a constant traveller, Nobel described himself as a 'man of no country'. When asked to write a biographical piece, he listed his principal virtues as keeping his nails clean and never being a burden to anyone. His principal faults included having no family, a bad temper and poor digestion. Alfred Nobel had a passion for orchids and horses. He was a voracious reader, a friend of Victor Hugo and an ardent admirer of the poetry and philosophy of Percy Bysshe Shelley. By nature he was depressive and melancholic, isolated and shy.

From the early days, when she nursed him through chronic ill health, Alfred Nobel loved and idealised his mother. Andriette lived to be ninety-four and Alfred remained always a devoted son. When he was eighteen, he fell in love with a Swedish girl, a pharmacists' assistant in Paris, only to suffer heartbreak when she died young of tuberculosis.

Nearly twenty-five years passed before love came again. At forty-three, Alfred Nobel advertised in Vienna as an 'elderly gentleman' requiring an intelligent and sophisticated personal secretary. The advertisement was answered by a thirty-three-year-old impoverished countess of great beauty and style, Bertha Kinsky, and within a week, Alfred dared to enquire whether her 'heart was taken'. She responded that it was and a week later, while Nobel was away on business, she fled to her lover. Nearly thirty years later, Bertha von Suttner was honoured with a Nobel Peace Prize.

Rejected by a woman of substance and mortified at his premature declaration of affection, Nobel's eye settled, a few months later, on Sofie Hess, a florist's assistant twenty-three years his junior. Sofie was uneducated, sometimes crass and increasingly demanding of luxury and funds. Alfred, who valued appearances, vacillated between affection and disdain. Over the fifteen years of their relationship, Sofie enjoyed her standard of living but resented Alfred's patronising and pedantic ways and chafed under his jealousy and restrictions. Unwilling to be allied with the socially inferior Sofie, Alfred complained about the lack of intellectual stimulation in her company and described to her as 'pushy'.

Sofie, who referred to Nobel as her 'Grumble Bear', divided her time between spas, her Paris apartment and her house near Salzburg. Nobel often addressed his correspondence to Madame Nobel, probably to alleviate any appearance of impropriety. She also increasingly used the title Mrs Nobel, leading to much confusion among friends and acquaintances.

When Sofie declared herself pregnant by another man in 1891, Alfred cut off their relationship and removed her from his will. He wrote to say that the one thing he could not forgive was someone making him appear ridiculous. The father of Sofie's child, an Austrian cavalry officer, married her – but in name only. After the ceremony, he kissed her hand then disappeared forever.

In the late 1880s, Sir James Dewar and Sir Frederick Augustus Able, members of the British government's Explosives Committee, were given access to secret documents outlining Nobel's discovery of smokeless gunpowder, ballistite. When Dewar and Abel patented their own version of smokeless gunpowder, Nobel, a man of great integrity, sued. The Nobel suit moved through the Court of Appeal to the House of Lords over three long and expensive years until, in 1895, the judgement went against Alfred Nobel. Although Nobel was ordered to pay court costs of £28,000, there was vindication of sorts in Mr Justice Kay's remarks: 'It is totally clear,' he said, 'that a dwarf who has been allowed to climb onto the shoulders of a giant can see somewhat farther than the giant himself.'

In 1893, after nearly a decade of crippling patent battles, industrial accidents, losses in the stock market and betrayal by his French partner, Nobel returned to his apartment in Paris. He was suffering from heart trouble – treated, ironically, with nitroglycerine. Alfred Nobel was increasingly depressed and referred to his 'despicable loneliness'. Newspapers of the time referred to him as the 'wealthiest tramp' in the world. 'More than most,' he wrote to Sofie, 'I have lived with the pressure of desolate loneliness.'

In December 1896, just four months after the death of his brother Robert, Alfred Bernhard Nobel collapsed at his desk and died a few days later of the effects of a stroke in his magnificent villa at San Remo on the Italian Riviera. His only companions were a doctor and his servants. With part of his brain affected, Nobel was able to communicate only in his native Swedish: the doctor was Italian and the servants were French, so no one was able to understand his last words. He died as he had feared, with 'No kind hand of a friend or relative to close my eyes and whisper in my ear a gentle and sincere word of comfort.'

By the time of his death at sixty-three, Alfred Nobel, who believed fervently that inherited wealth was a misfortune that encouraged laziness, had changed his will for the third and last time. In this final, handwritten, document, smaller personal bequests replaced the original twenty per cent of his estate originally designated for his nieces and nephews. Instead, the bulk of his fortune was directed to the establishment of a foundation to fund a series of awards, including the Peace Prize. Three of his brothers' children challenged the will and some settlement was made. For Nobel, there was clearly some satisfaction in withdrawing the legacies from his family, a sentiment expressed when he wrote to Sofie that he took great delight in advance 'in all the widened eyes and curses the absence of money will cause'.

Nobel's assets were spread over eight countries and included the armaments factory at Bofors in Sweden. After debts and claims had been settled, the Nobel Foundation was funded with just under $7 million. Today, the Foundation is worth billions. Sofie, who had continued to write to Nobel asking for money, produced more than two hundred of his letters after his death, letters he had asked her to destroy. The Foundation met her demands and purchased the correspondence, keeping it under lock and key for more than fifty years. Sofie retained some letters, thinking that she could extract further funds at a later date, but her ploy failed. As part of the agreement, she was paid an annuity for the remainder of her life.

Nobel's poetic side was embodied in *Nemesis*, a play he published just before his death. It was not well-received and the family, believing that the play would damage his reputation, had all but three copies destroyed.

At his death, Alfred Nobel held more than three hundred and fifty patents. The fundamental principles underlying Nobel's explosives became the foundation of solid-propellant rockets used, for example, in the space programme.

The big bang
SAM COLT

Born: July 19, 1814, Hartford, Connecticut
Died: January 14, 1862, Hartford, Connecticut

However inferior in wealth I may be to the many who surround me, I would not exchange for there [sic] treasures the satisfaction I have in knowing I have done what has never before been accomplished by man.

Sam Colt

YOU NEVER KNOW HOW THINGS WOULD HAVE TURNED OUT under different circumstances, but there is no question that early trauma and loss lit a fire in Sam Colt. Left motherless at seven and sent to a foster home, a life that had been secure and happy disappeared overnight. Not all of Colt's siblings weathered the change as well as he did: his sister committed suicide and a brother was convicted of a gruesome hatchet murder.

Sam's early loss resulted in an independent spirit and a driving ambition to succeed. A fertile imagination projected him out of the sad circumstances in which he found himself. At ten, Sam, a budding inventor and born promoter, went to work in his father's silk mill, where he became familiar with machinery and production techniques. At fifteen, he posted a public notice to the effect that he would blow a raft 'sky high' on Ware Pond on July 4. Colt's device missed the raft, drenching the crowd, dressed in their finery, with pond water and mud. On another occasion, Sam organised a pyrotechnic display that set his school on fire.

Perhaps out of concern for the course that events were taking, Charles Colt arranged an apprenticeship for his sixteen-year-old son on a ship bound for India. It was on that trip that young Sam almost certainly became acquainted with the Collier repeating firearm used by the British Army. The Collier chamber was rotated by hand between shots. In Colt's later design, the cylinder rotated automatically when the hammer was cocked.

Back in America, Sam Colt's father agreed to finance a prototype of the crude wooden pistol with a revolving chamber that Sam had carved on the journey back from India. John Pearson, a fine gunsmith, was hired to create a version in metal, but found dealing with Sam Colt very trying, not least because the inventor was careless about paying his bills. 'Make your expenses as lite as possible,' Colt wrote to Pearson. 'Don't be alarmed about your wages, nothing shal be rong on my part, but doo well for me & you shal fare well.' (Colt adhered to the adage that anyone who spelled a word the same way more than once had no imagination.) Pearson, who eventually removed himself from Colt's operation, later claimed that he was the true inventor of the repeating pistol.

To raise the capital necessary to continue development of his revolver and finance a patent, the celebrated 'Dr Coult of New York, London and Calcutta' took to the road, offering laughing gas for fifty cents a sniff at the local fairs. It didn't take long to raise the necessary funds. In 1836 with US Patent No. 138 in hand, twenty-three-year-old Samuel Colt approached a group of backers for the $200,000 necessary to open his first firearms factory, The Patent Arms Manufacturing Company, based in Paterson, New Jersey.

Sam spent lavishly in Washington, where he lobbied government officials and military figures, drank heavily, and fell into difficulty with creditors. In spite of his efforts, the Army rejected the five-shot revolver in favour of what it already had, and West Point rejected the new pistol as 'too complicated'. The few orders that came in were not enough to stave off bankruptcy. In 1842 the Arms Manufacturing Company closed its doors.

In a remarkable instance of self-inspection, Colt was wise enough to understand that a different approach had to include a new personal style. Penniless, he moved to New York where he started an aggressive course of self-improvement that included the study of chemistry and engineering. Colt also joined the Historical Society, where he mingled

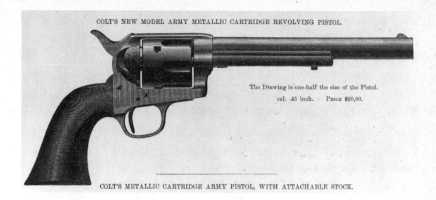

COLT'S NEW MODEL ARMY METALLIC CARTRIDGE REVOLVING PISTOL.

The Drawing is one-half the size of the Pistol.
cal. .45 inch. · PRICE $20,00.

COLT'S METALLIC CARTRIDGE ARMY PISTOL, WITH ATTACHABLE STOCK.

with other scientists and inventors, including Samuel Morse, with whom he undertook some experiments. To his inventive imagination and driving ambition, Sam Colt added credentials. Colt the outsider became Colt the insider.

Other types of rifles and revolvers, as well as waterproof ammunition, underwater mines detonated by a telegraphed signal, and an underwater telegraph cable were added to Colt's list of inventions. The government paid Colt $50,000 to demonstrate his underwater mine by blowing up a 500-ton schooner on the Potomac River. The guests, including President Polk and members of Congress, enjoyed the successful demonstration, but not enough to award a contract. John Quincy Adams dismissed Colt's invention as an 'un-Christian contraption'.

The tide turned with a letter from war hero Captain Samuel Walker of the Texas Rangers. 'In the summer of 1844,' Walker wrote, 'Col. J.C. Hayes, with fifteen men, fought about eighty Comanche. Without your pistols, we should not have had the confidence to undertake such daring adventures.'

The Comanche indians could fire twenty arrows a minute, while the conventional pistol only managed one or two rounds in the same amount of time. The Colt revolver tipped the balance. Walker and Colt became good friends and, together, redesigned the revolver, resulting in an order for a thousand of the new six-shot, .44 calibre 'Walker' pistols at a cost of $25,000. Sam Colt was off and running. In 1849, Colt's patent was renewed for seven years and a manufacturing facility was set up in Hartford, Connecticut in a factory belonging to Eli Whitney, Jr. In less than a decade, Sam Colt had become a millionaire.

Sam Colt's fierce independence showed through in this letter. 'I am working on my own hook now,' he wrote, 'and I have sole control and management of my business and intend to keep it as long as I live without being subject to the whims of a pack of damn fools and knaves styling themselves as a board of directors.'

Colt hired Elisha King Root, one of the master machinists in the Connecticut River Valley, to oversee production. Root transformed the Colt factory into a sophisticated mass-production facility that could push out guns at lower unit cost. Colt's perfection of the assembly line and introduction of a system of interchangeable parts paved the way to America's industrial revolution. Colt was also among the first industrialists to manufacture product overseas when he set up a factory in England in 1853. He would, he claimed, supply the English government with the best 'peesmakers' in the world.

The American Civil War and the Crimean War made the years between 1848 and 1865 one of the biggest growth periods for armaments, and Colt was a major beneficiary. As a businessman, he was not averse to selling firearms to both the Native Americans and the forces fighting against them. By 1852 Colt was a national celebrity and one of the wealthiest businessmen in America, overseeing his armoury from a chair at his roll-top desk while he puffed away on Cuban cigars.

'God made all men,' went one popular saying. 'Sam Colt made them equal.' Sam Colt wanted to be famous. He created a cult around himself and branded his product in a way that few, if any, had done before him; to this day the word the French generally use for a revolver is 'un Colt'. Armed with his newfound wealth, Colt mingled with the rich and famous in Newport, Rhode Island, where he met Elizabeth Hart Jarvis, the daughter of an affluent and socially prominent family. They married in 1856 in an extravagant wedding that featured a steamboat and a rifle salute. Elizabeth and Sam were a devoted couple and forged a strong partnership. Home for the Colts was Armsmear, a Hartford, Connecticut mansion with two dozen rooms, a conservatory and sculpture-filled gardens, but the thrust of their interest was philanthropy. Elizabeth carried on this vision during her forty-three years of widowhood.

Things were going well at last, but not for long. During five years of marriage, Elizabeth and Sam lost four children, three to illness and one stillborn. Suffering from gout and rheumatism, Colt was under a great deal of stress with the advent of the Civil War and the demands that supplying both the North and the South put on his factory. He died in 1862 at the age of forty-seven, having never fired a gun at another person. The Colt Armoury stood quiet as fifteen hundred workmen filed past his casket. Sam Colt's only surviving heir was his three-year-old son, Caldwell, the only child to survive infancy.

Colt left Elizabeth one of the wealthiest women in America, with a fortune topping $15 million, but that didn't compensate for her great personal losses. Within a period of five years, Elizabeth had lost her husband and four children. When she was seventy-two, her one remaining son, Caldwell, contracted tonsillitis and died. Other than his wife and his son, the only other beneficiary of Sam Colt's estate was his brother John's son, Samuel Caldwell Colt.

Two years after Colt's death, Elizabeth stood at her window and watched as the Colt armoury in Hartford burst into flames and burned to the ground. The great blue dome with its symbol of the rampant colt in blue gilt zinc crashed dramatically into the inferno. The dome would be rebuilt and the new factory would continue to produce firearms for another hundred years.

Disappearing act

BETTE NESMITH GRAHAM

Born: March 23, 1924, Dallas, Texas
Died: May 12, 1980

In a discovery there must be an element of the accidental,
while an invention is purely deductive.

Thomas Edison

SCHOOL WAS NOT A GREAT SUCCESS FOR BETTE CLAIRE MCMURRAY. A fractious, strong-willed student, Bette's teachers may well have breathed a sigh of relief when she finally dropped out of school at seventeen, in the midst of the Second World War.

In spite of the fact that she couldn't type, Bette managed to land a job as a secretary at a law firm. On a wing and a prayer, they sent her to secretarial school, where she gained her high-school diploma. Within two years, Bette had married Warren Nesmith and they produced a son, Michael. Warren went off to war and, when he returned, the couple divorced. For the next twenty years, Bette was a single working mother.

Bette did quite well, working her way up to a position as executive secretary with Texas Bank & Trust, but her typing skills still left something to be desired. When IBM came out with the new electric typewriter, complete with carbon film ribbon, Bette and a lot of her co-workers found to their dismay that their erasers were leaving large dark smears. When Bette decided to earn a little extra money dressing the bank windows for the holiday, she noticed that when the display artist made a mistake in his lettering, he simply painted it over.

'I put some tempera waterbased-paint in a bottle,' she wrote later, 'took my watercolor brush to the office and used that to correct my typing mistakes.' Bette used her paint and paintbrush to correct typing mistakes for the next five years.

When Bette's co-workers asked for bottles of their own, she realised that she might be on to a good thing and decided to take it further. One of her son's chemistry teachers gave some advice and a local paint manufacturer taught her how to grind and mix pigment. Turning the kitchen into a laboratory and her garage into a factory, Bette scraped together $200 to pay a chemist to develop a fluid with a solvent base to replace her water-based product, thus perfecting a quick-drying solution that was virtually undetectable. Bette's son Michael and his friends filled literally hundreds and then thousands of bottles with the magic fluid, using squeezable ketchup dispensers. The original name, 'Mistake Out', was replaced with 'Liquid Paper'. In 1956, Bette applied for a trademark. In the ten years it took for Bette's company to become profitable, she continued to work as a secretary at IBM to make ends meet. Her career came to an abrupt end when she inadvertently typed 'The Mistake Out Company' instead of IBM at the bottom of a business letter. She was fired on the spot.

In 1962 Bette married Bob Graham, a salesman who provided the marketing expertise that she needed to convert her invention into a money-making proposition. Soon they had built up a network of dealers and office supply stores. Articles and ads appearing in trade publications boosted sales until, in 1968, Bette's company was grossing more than $1 million a year. By 1975, Bette employed two hundred people, produced twenty-five million bottles of Liquid Paper annually and distributed her product in thirty-one countries. When Bette Nesmith offered her company to IBM, however, they turned it down.

In 1979, the Gillette Corporation offered Bette $47.5 million for Liquid Paper, which had annual revenues of $38 million. In addition to a lump sum payment, she received royalties on every bottle sold through to the year 2000. Ironically, Gillette bought Liquid Paper primarily for the typewriter ribbon division, believing that correction fluid would cease to be important in the upcoming computer-driven market. What happened was exactly the opposite, and Gillette paid off its investment in three years.

When Bette died in 1980, she left half her fortune to her son, Michael, a member of the pop group The Monkees, and half to her favourite charities. Michael invested some of his inheritance in Pacific Arts Studio, a forerunner of the modern video music production studios.

Tunnel vision

MARC ISAMBARD BRUNEL

Born: April 25, 1769, Hacqueville, France
Died: December 12, 1849, London, England

I do not think there is any thrill that can go through the human heart like that felt by the inventor as he sees some creation of the brain unfolding to success ... Such emotions make a man forget food, sleep, friends, love, everything.

Nikola Tesla

AS THE REVOLUTION SPREAD INTO THE FRENCH COUNTRYSIDE, Marc Isambard Brunel, architect and engineer, caught a steamer bound across the Atlantic. Landing in New York in 1793, he was never again to live in his native France. Brunel was soon exploring the wild country around the St Lawrence River, envisioning canals that would link the Great Lakes with the Hudson River and enjoying his appointment as New York's Chief Engineer.

Alexander Hamilton, one of America's Founding Fathers and first secretary of the Treasury under George Washington, numbered among Brunel's friends. At a dinner party in 1798, a group of men, among them Marc Brunel, gathered around Hamilton's dining table. Candles flickered as they lit their cigars and settled back for a brandy. Had they heard, asked a fellow countryman of Brunel's, about the cumbersome and costly way in which the English were fashioning the blocks for their ships? Hamilton and Brunel, both worldly and intensely curious, listened carefully. Whatever the English, a major naval power, did – or didn't do – was worth noting. Without leaving the table, and knowing little about ships, Brunel leaned forward in his chair and threw out a suggestion for improving the process. The men nodded approvingly. That idea, sketched on a napkin, blossomed into the design for Brunel's renowned block machinery, which he patented in 1801. The English firm that held the monopoly for ships' blocks contemptuously turned down Brunel's idea, as did all the great British shipyards, but Earl Spencer, First Lord of the Admiralty under Prime Minister William Pitt, was quick to appreciate the advantages of the design. Brunel's proposal was eventually accepted, but manufacture proved difficult.

Soon after that dinner, Brunel left for England. In part, he was excited about his idea; in part, he was anxious to reunite with Sophia Kingdom, an Englishwoman whom he had met in France. They married and had three daughters and one son, Isambard Kingdom Brunel.

Marc Brunel had the soul of an inventor. Take, for example, the evening he stood at Earl Spencer's card table and watched as one of the players experienced difficulty shuffling the cards. Within a few days, Brunel reappeared at Althorp, delivering to Lady Spencer a small shuffling box requiring only the turning of a handle to mix the cards. He was, wrote his granddaughter, a pure engineer trying to give people whatever it was they wanted at the time.

Translating invention into a sound business was more difficult for Brunel and much of that was just bad luck. He devised a wide variety of saws, including the circular saw, to work the oak England imported for its ships, and set up sawmills employing equipment of his own design. That factory was destroyed by fire. When Brunel noticed that soldiers were returning from the battlefields in Europe in 1812 with their feet bleeding and torn, he turned his attention to footwear. Before long, he had invented machinery capable of turning out four hundred pairs of durable and inexpensive shoes a day in nine different sizes.

When Napoleon escaped from Elba and the war with France resumed, the British government placed a large order for shoes for the troops. When the Battle of Waterloo reversed the course of the war, the order was rescinded and Brunel, unprotected, was left with eighty thousand pairs of shoes that nobody wanted. There was no recourse. Marc Brunel was consigned to Southwark debtors' prison for two months. His wife, as was the custom, accompanied him. Friends kept an eye on him, and when Russia made overtures to Brunel, Lord Spencer intervened: 'However lacking Brunel may be in commercial instincts, his peculiar genius for invention is too good to lose.' Many eminent scientific figures were in agreement.

Marc Brunel believed wholeheartedly that conundrums usually solve themselves if you know where to look. The year was now 1822, and for months Brunel had been turning over in his mind the way in which a tunnel might be constructed underneath the Thames at Rotherhithe, at which point nearly four thousand passengers a day were being ferried across the water. At a time when large-scale engineering projects were in their infancy, most people had proclaimed such a tunnel 'impracticable'.

Construction of the tunnel under the Thames at Rotherhithe.

Walking through the shipyard one day, Brunel picked up a bit of timber and found himself admiring the way in which the mollusc *Teredo navalis* bored its way through the hard wood using a pair of strong shell valves to forge the way. Before long, Brunel had translated those valves into an iron shield, which he patented in 1818. Behind that shield, thirty-six men could bore their way through the soft ground under the Thames, each man isolated in an independent cast iron cell. Bricklayers followed immediately behind. The excavated soil was passed back through the tunnel, much as the mollusc passed its wooden meal through its digestive tract.

In 1825, Brunel laid the first brick for the tunnel project. His son Isambard, now twenty, laid the second. During the more than eighteen years that ensued, the Brunels suffered financial hardship and mental strain as one of the world's greatest engineering projects to date unfolded.

In 1827, Isambard Kingdom Brunel assumed the duties of chief engineer under his father for the Thames Tunnel project. This was an ambitious apprenticeship for the twenty-one-year-old who now found himself managing sophisticated engineering operations, mediating labour disputes, and overseeing the safety and well-being of his men. But Isambard Kingdom Brunel was a gifted engineer, a decisive leader and a man of great personal charm who quickly grew accustomed to being in charge. Isambard made his home in a little cabin that he fitted out for himself at the top of the tunnel shaft, where he would sit and smoke or talk with friends, catching a few hours of sleep sitting on a chair before the fire. For the first time, he was earning his own salary.

Progress on the tunnel was slow and arduous. Influxes of water overwhelmed and sometimes injured the men. One incident nearly cost Isambard Kingdom Brunel his life and was so costly to repair that the project was halted for seven years. When it opened again, Isambard, displaying his bent for showmanship, arranged a dinner for fifty dignitaries in the half-finished tunnel. Underneath the Thames, embraced by crude brick arches, people dined at tables in the glow of candelabras fuelled by the new 'portable gas' to music provided by the Coldstream Guards.

In 1843 the great Thames Tunnel was completed. The first visitor, handed through from hand to hand, was Isambard the third, just three months old. In the course of the first two days, fifty thousand people paid a penny each to walk through the tunnel. Queen Victoria was the first of two million visitors in the first year.

From an engineering perspective, the Thames Tunnel was a breakthrough;

financially, it was a disaster. Not even enough money remained at the end of the project to build the simple ramps that would allow vehicular traffic. In 1866, the tunnel was purchased by the East London Railway and opened to traffic. Eventually, this great engineering masterpiece was incorporated into the London Underground System.

Marc Isambard Brunel was paid £5,000 for the use of his patent at the start of the project but was never to receive the additional £5,000 contractually owed to him; £1,700 was all the shareholders could manage. He was knighted by the Queen and lived a comfortable and sedate life until he died at the age of eighty.

Isambard Kingdom Brunel on board the Great Eastern. *An hour later, Brunel suffered a stroke.*

The biggest and the best
ISAMBARD KINGDOM BRUNEL

Born: April 9, 1806, Portsea, England
Died: September 15, 1859, London, England

*The history of invention records no instance of grand
novelties so boldly imagined and so successfully carried
out by the same individual. Brunel was less successful
when he was less bold. When he could not be grand,
he was nothing at all.*

The *Morning Chronicle*, London, on the occasion of Brunel's death

Brunel remained all his life in love with the impossible.

Kenneth Clark, *Civilisation*

MARC ISAMBARD BRUNEL, THE MASTER ENGINEER behind the Thames
tunnel, rejoiced in his gifted son, Isambard Kingdom Brunel. Isambard learned to
draw at the age of four and was familiar with Euclid by the age of eight. At seventeen,
Isambard went to work on his father's tunnel project. In his professional journal, using
a shorthand system to protect his privacy as he recorded thoughts and feelings
Isambard reported that he was very short of money and described himself as conceited.
'I am looking forward to the gaz engine,' he wrote. 'I am building castles in the air
about steamboats going fifteen miles an hour; going on tour to Italy; being the first to

go to the West Indies; making a larger fortune and building a house for myself.' Within thirty years, most of Isambard Kingdom Brunel's daydreams had come true.

Isambard's promotion to chief engineer for the tunnel project in 1827 was a big step up. The heady connection with influential men and learned societies accruing through his father's reputation opened doors for him at a young age. Ever after, Isambard found it difficult to accept a situation in which he was not in charge. A decisive and charismatic figure, at best Isambard ignored, and at worst rebelled against, the imposition of authority. Budgets and shareholder concerns took a back seat to creativity.

In 1829, while recuperating from a bad accident in the tunnel, Isambard responded to an advertisement inviting engineers to present plans for the Clifton Bridge in Bristol. The exquisite designs he submitted illustrated a suspension bridge 870–914 feet in length and 240 feet high spanning the Avon Gorge. The span was suspended by great chains from massive towers guarded by two Sphinxes. His plans were accepted after much debate. 'I produced unanimity,' Brunel wrote home to his parents, 'amongst 15 men who were all quarrelling about the most ticklish subject – taste.' The bridge, begun in 1836, was abandoned twenty-three years later due to lack of funds, only to be completed after Brunel's death.

When twenty-six-year-old Isambard entered his design for the railway line between Bristol and London, plans that included a bold two-mile tunnel through Box Hill near Bath, the committee turned him down. They decided instead on the plan that cost the least. Brunel, who had a compelling vision for fast, long-distance travel, wrote a strong letter arguing against the decision. The committee reversed its decision, and Brunel was appointed engineer by a majority of votes. He made it clear that he would not participate if economic considerations were paramount.

In order to complete an examination of the proposed line and oversee his crew, Brunel contracted for a special horse-drawn carriage known as the 'flying hearse' in which he lived and worked for the duration of the survey. Working twenty-hour days, Brunel commented that the survey had been 'harder work than he liked'. When the survey was completed, on schedule, Brunel retired for twenty-four hours with instructions to remain undisturbed. When his brother-in-law looked in, Brunel was lying with a complete cigar in ash spread across his chin.

The tunnel through Box Hill was controversial. Many nervous passengers opted to leave the train, circle the hill by road, and meet the train at the other end. Just as revolutionary as the tunnel through the hill was Brunel's recommendation to make the rail gauge seven feet rather than the traditional four feet plus favoured by George Stephenson. A wider gauge, he argued, would allow them to lower the centre of gravity of the rolling stock and achieve greater speed and a better ride.

When Brunel's new line inevitably met up with the older lines constructed to different specifications, the 'Battle of the Gauges' was entered. Brunel – imaginative and daring – found himself pitted against his friend, George Stephenson's son, Robert, who was conservative by nature. The debate reflects an evolutionary stepping stone in the world of invention, what Jonas Salk described later as the battle between vision and stasis. Claiming that a faster train service was 'unnecessary', and anxious about funds, Parliament passed an Act establishing the narrow gauge with iron rails laid over stone blocks as the regulation for England, in part because hundreds of miles of narrow gauge already existed. A writer in 1935 referred to this as 'the supreme example of the stupidity of uncontrolled development'.

Brunel's vision projected further into the future than most people were able to go. Fortunately, he was able to project that vision in a charismatic and persuasive way. Isambard Kingdom Brunel was not only an ideas man, but a salesman, a man who made a strong impression on all he met.

At twenty-nine, Isambard Kingdom Brunel was well on his way, enjoying phenomenal success, earning good money and living in a new house at 18 Duke Street with his wife, Mary Horsley. Although the family lived in increasing luxury, Brunel himself maintained an austere existence. A devoted husband and father in spite of long periods away from home, he worked into the early hours and seldom went out except with his family or to go to a learned meeting.

The first train, travelling at thirty-six miles an hour between Paddington and Maidenhead in 1838, was pulled by the elegant engine *North Star*, manufactured in Robert Stephenson's workshops in Tyneside. When the fanfare died down, however, troubles began to appear, many of which were attributed to Brunel's determination to lay the tracks on timber piles rather than stone blocks. Isambard stood his ground, seemingly unperturbed by the anger and anxiety swirling around him.

Railroads required bridges, to which Brunel could apply his full imagination. The first was over the Thames at Maidenhead. Brunel also devised a system of disc and crossbar signals used on the broad gauge railway line and took an intense interest in Wheatstone's Electric Magnetic Telegraph, which he installed on the line between Paddington and West Drayton in 1939. Queen Victoria, who with her husband Prince Albert was open to new things, had an apparatus built into the roof of her private coach by means of which she could signal 'stop', 'slow', and 'speed up'.

Brunel nearly collapsed under the strain of the Great Western Railway project and the bitter criticism aimed at him when things didn't work to plan. A group of shareholders from Liverpool called for his replacement, but the other directors held firm. The controversy died down when the line between Bristol and London was opened in 1841. With the project complete, Brunel issued a challenge. "Why not make this line still longer," he urged, "with a steamboat connection between Bristol and New York?"

"By gad," came the reply, "the young fellow will be taking us to the moon."

The *Great Western*, an oaken paddle steamship with a single funnel, was built to Brunel's design in Bristol, with the intent of crossing the Atlantic. Brunel had no previous shipbuilding experience. The maiden voyage began on April 8, 1838 and was completed fifteen days later in New York.

Brunel next proposed an iron ship, with water-tight compartments and a screw propeller. The 322-foot *Great Britain*, completed in 1843, was one third again longer than any Navy battleship and had sixty-four staterooms. Although she went aground off Ireland, the ship proved to be a great success.

Sometimes invention came in useful closer to home. Playing with his son, Brunel pretended to swallow a sovereign and bring it out through his ear. Unfortunately, the sovereign lodged in Brunel's windpipe. When an operation failed, Brunel designed an apparatus onto which he was strapped and swung head downwards. The obstruction shifted, someone struck Brunel between the shoulders and the coin dropped from his mouth. When Macauley rushed into the Athenaeum Club yelling 'It's out!' no one needed to ask what he was talking about.

Brunel's inventiveness took a completely different turn when Florence Nightingale, nursing at the front during the Crimean War, alerted the British government to the deplorable hospital conditions. Brunel was approached to design a model hospital with all modern conveniences. His sketchbook is filled with drawings of invalid baths on wheels, ventilation and drainage systems that improved sanitation, and small boilers. Less than three months later, the first consignment of prefabricated materials for free-standing clinic buildings was despatched to Turkey.

There is almost an inevitability about the end of the Brunel story, given the inventor's insatiable energy for cutting-edge technology, his disdain for conservative fiscal restraints, and his high degree of self-assurance. Just at the height of his personal and professional success, forty-year-old Isambard Brunel set out on a course that was to prove disastrous.

The stampede to New South Wales in response to the discovery of gold provided Brunel with the perfect excuse to design a third ship, one of unprecedented size that could go around the Cape to Australia loaded with goods and people. Brunel turned an impatient ear to those who proposed a more moderate vessel. His drawing board was covered with sketches of a great double-hulled iron ship capable of transporting four thousand passengers. Eight engines and six masts powered the ship; ten anchors kept her in place. Brunel admitted that he had staked his reputation – not to mention the stockholders' money – on the *Great Eastern* project. He also admitted that he would need to bear, 'solely, and very heavily, the blame of failure'.

A vast wall of iron rose up on the southern end of the Isle of Dogs. The double-hulled eighteen-thousand tonne *Great Eastern*, six times larger than any existing vessel, called for unprecedented production techniques. None of the traditional shipbuilders was conversant with the process of riveting wrought iron plates; none of the experienced captains had the skill to handle a ship of this size and description. To complicate matters, the project was hopelessly undercapitalised.

The inventor was less worried about construction than he was with the problem of moving the *Great Eastern* from the hard into the water. When directors, anxious for revenue, issued three thousand tickets, sightseers thronged the pier and private boats blanketed the river, refusing to retire to a safe distance. Eminent engineers and scientists from around the world, among them the King of Siam swathed in gold robes, gathered to watch as the behemoth was lowered into the Thames. Circumstances couldn't have been more stressful.

At noon, Brunel, standing on a platform in the ship's stern, gave the sign. Lighters on the river began to haul on the great chains, hydraulic presses pushed. A great rumbling arose from the hull. 'She moves,' someone cried out.

The stern slipped forward, ahead of the rest of the ship. Five men were thrown

The Great Eastern *under construction in the Isle of Dogs*

overboard when the windlass spun out of control. It was only with great force that the ship's progress was arrested. Aware of the danger to those around, Brunel gave the order to restrain the ship's progress towards the water, but in so doing he interrupted the process at a critical point. When the launching was resumed that afternoon, the great ship refused to move. The mob swarmed over the dock. Brunel described the events as 'cruel'.

A month later, they tried again. The ship refused to budge, giving only an inch here and there as increased pressure was applied. Just as things seemed to be going well, a dense fog rolled in. The *Great Eastern* lay like a beached whale just feet away from the water. Public and private commentary swung against Brunel, taunting him and referring to 'monuments to his vanity'.

Saturday January 30, 1858 dawned blustery, with rain and a bad southwesterly gale blowing hard against the ship's broadside. Captain Harrison advised that it would be 'next to insanity' to attempt to float her in such conditions, but by Sunday the *Great*

Eastern was safely moored in the Thames. In a letter, Brunel advised his son 'never to underestimate the advantage of prayer'.

The launch had overrun estimates by more than £100,000. With the company bankrupt and Brunel himself considerably poorer, the *Great Eastern* hull lay idle in the river for more than a year. Brunel went back to work, but he was a changed man, the continued brunt of sarcasm and reproach. The store of energy that had led him on his spirited life's journey was gone. The last few months of his life were devoured with a passion for his ship, which had been taken over by the Great Ship Company in 1858.

As the *Great Eastern* prepared to set sail for America, Isambard Kingdom Brunel supervised every detail. A few days before the voyage, a representative from the London Stereoscopic Company came on board to take some slides, including one of Brunel standing beside a smokestack. Less than an hour after that image was recorded, Brunel suffered a stroke and was taken home to Duke Street, paralysed.

On September 9, 1859 the *Great Eastern*, with bands playing and flags flying, left the dock. Suddenly, a terrible explosion rocked the stacks, due entirely to the careless failure to open a stop cock. Parts of the deck, cabins and all manner of debris were thrown into the air. Five men were killed. With news of the disaster, Isambard Kingdom Brunel gave up on life. He died on September 15, 1859 at the age of fifty-three, a man whose ideas had crushed their creator.

An article in the *Morning Chronicle* applauded Brunel, going on to say that he was the right man for the nation and the wrong man for the shareholders. 'Those must stoop who would gather gold,' they wrote, 'and Brunel could never stoop.' Brunel's restless mind went on to new projects before the ones before were finished, dimming any hopes of profit.

After a short history of voyages and mishaps, the *Great Eastern* was auctioned off for £25,000 to the Telegraphic Construction Company and used to lay the first trans-atlantic telegraph cable. When she was finally broken up, the bodies of two workmen were found in the three-foot space between her two hulls. Maybe, people said, this accounted for her bad luck.

Isambard Brunel was opposed to government intervention in commercial enterprises, particularly in respect to patent law. In the eighteenth and early nineteenth centuries, the patent system was, by virtue of being expensive, inaccessible to the workman whose creativity gave rise to a potentially grand idea. A case in point is Thomas Highs, a craftsman in the mid-1700s who designed the prototype of a spinning jenny, only to lose all rights and kudos to Richard Arkwright.

While others clamoured for patent reform, Brunel, viewing patents as a check on enterprise, voted for abolition. His father had made use of the patent system to protect his investment, as had Brunel's friend Robert Stephenson. Isambard Kingdom Brunel refused to take out patents. In his view, the 'dream of thousands' in profit held creative genius back. Better, he said, to earn a few pounds more often with good ideas than to spend all one's time on one grand scheme. Brunel argued that the patent system rarely, if ever, benefited the genuine inventor. He seemed not to care when others borrowed or even patented ideas to which he could legitimately lay claim.

Echoing Jonas Salk's view that invention is the modern-day form of Darwinian evolution, Brunel saw the human process in innovation. 'I believe,' he wrote, 'that the most useful and novel inventions and improvements ... are mere progressive steps in a highly-wrought and highly-advanced system otherwise known as evolution.'

When Brunel died, his finances were greatly reduced, due in large part to the *Great Eastern* adventure. An inventory was demanded on his Duke Street house and its contents. His total estate was valued at £90,000.

Taking a dip
JOHN PHILIP HOLLAND

Born: February 24, 1841, Liscannor, Ireland
Died: August 12, 1914, Newark, New Jersey

An inventor is a kind of beggar who seeks support for
patentable ideas, and a beggar cannot be a chooser ...
Since he is, by nature, a dreamer, he often comes to the
practical test with no financial resources of his own.

Richard Knowles Morris, *John P. Holland, Inventor of the Modern Submarine*

THE HATCH COVER POPS OPEN AND A BESPECTACLED MAN wearing a bowler hat peers out at amazed spectators. Meet John P. Holland, looking more like the schoolteacher he once was than the inventor of a newfangled boat that disappears beneath the surface of the water, only to emerge later with all occupants safe and sound.

Born in a small Irish village in County Clare, Holland's beginnings were inauspicious. At the age of twelve, just after his father died, he joined the Order of the Irish Christian Brothers, taking his initial vows in June 1858. He would, he decided, become a teacher. But by the time he was thirty-two, without any family ties remaining in Ireland, John Holland withdrew from the Irish Christian Brotherhood, packed his bag and sailed, steerage class, for America. Once in Boston, he sent his early designs for a submarine to the Department of the Navy, receiving in return the comment that his ideas were 'impractical'.

If the American government wasn't interested, the Irish were. The 'Fenian Skirmishing Fund' set up by the Irish in America to promote their country's independence, was ready to fund Holland's prototype submarine. On May 22, 1878, a team of stallions backed Boat Number 1 – fourteen and a half feet long, with a three-foot beam – into the Passaic River, off Paterson, New Jersey. Spectators held their breath as the small craft bobbed on the surface for a moment before disappearing and settling on the river bottom.

In spite of this disappointing start, by June the little submarine was ready for her next public appearance. John Holland, wedged into a space only three feet wide, three feet eight inches long and a little more than two feet high, took her down for a successful dive. On the bank, hidden behind the bushes, a man with binoculars – reputed to be the British spy Major Henri LeCaron – watched and waited.

The Skirmishing Fund was sufficiently impressed to commission a second, larger submarine, the *Fenian Ram*. When thirty-seven-year-old John Holland gave up his teaching career to concentrate full time on this project, he said that his biggest regret was leaving his students, 'the only ones who didn't think I was crazy'.

The *Fenian Ram*, thirty-one feet long with a six-foot beam, reflected Holland's view that a submarine should look and behave like a porpoise. He remarked that it seemed uncommonly difficult for other designers to adopt this sleek form, probably, he speculated, because the powers that be were reluctant to give up the decks on which they were accustomed to strut. The *Ram* was launched in June 1881 carrying projectiles designed by John Ericsson, designer of the *Monitor,* a revolutionary armoured ship used during America's Civil War.

When the first projectile was released from the Ram's rotating turret, it rose more than sixty feet in the air before falling back into the water and coming to rest in the muddy riverbed. Some adjustments were made and a second projectile was released. Travelling at double the speed of the first, it also rose sixty feet into the air, cleared the breakwater, and hit a piling. 'What's that?' yelled a surprised fisherman, popping up from behind a rock.

Testing of Holland's submarine continued. Seeking deeper water, the *Fenian Ram*, sealed for a dive, made its way down New York Harbour, John Holland at the helm. Suddenly a strange scratching was heard from above and Holland noticed that his view was obscured by what appeared to be rags flapping in the breeze. Opening the hatch, the crew discovered a young boy hanging onto the turret for dear life. It appears that the prospect of going below was even more scary, so the crew headed gingerly for shore with their young passenger still clinging on. On another occasion curiosity almost killed the cat when the engineer decided to take the *Ram* out for a run on his own. When water splashed in through the open hatch, the hapless engineer suddenly exploded through the turret, propelled by the air escaping from the boat.

Dissension in the ranks of the Fenian Brotherhood spelled the end of their support for John Holland and his submarine, leaving our inventor with no practical support. Between the years 1889 and 1893, Holland worked as a draftsman for $4 a day.

Then, in 1893, Holland entered a government competition under the umbrella of the John P. Holland Torpedo Boat Company, whose backers had raised money to fund a prototype. Holland was appointed manager at a salary of $50 a year. Elihu B. Frost, an influential lawyer in Washington, was Secretary/Treasurer. In the company reorganisation that followed the award of a government contract, John Holland was given stock, but not a controlling interest. All of Holland's submarine inventions and patents would henceforth be the property of the Torpedo Boat Company.

The marriage with the Navy was not a happy one, and frustrations grew as the bureaucracy imposed an increasing number of unrealistic requirements on Holland and his team. Seeing that the government project was moving slowly, the Torpedo Boat Company secretly went ahead with a sixth submarine, the *Holland VI*, in 1896.

Announcing that he 'planned to take a dip,' Holland pulled away from the shipyard and was towed to Arthur Kill, off Staten Island. On St Patrick's Day, 1898, at 2:30 in the afternoon, the five members of the *Holland VI* crew slipped through the hatch. Just as John Holland closed the turret, the rain stopped and a brilliant rainbow stretched across the sky over Staten Island. The submarine gave a perfect performance.

When German entrepreneur Isaac Rice stepped onto the *Holland VI*, John Holland had no way of knowing that he was looking at a drastic change in his fortunes. All seemed well, and it was a great relief when Rice, impressed by what he saw, declared himself ready to put up money for necessary alterations to the *Holland VI*. But Rice and E.B. Frost had other ideas. Within a year, the Torpedo Boat Company had become a subsidiary of Rice's Electric Boat Company and John Holland found himself increasingly sidelined.

'You need a rest,' his partners said, as they encouraged Holland to take a holiday abroad. With Holland safely on board ship, Isaac Rice and Elihu Frost wrote a cheque for the back taxes on Holland's European patents, providing a lien and virtual control. John Holland, like King Gillette (see page 3), had put himself into the hands of businessmen far more wily and ruthless than he. Holland was demoted once again, this time to Chief Engineer. The friends and associates he had trusted had virtually stripped John Holland of his rights. No one went to Holland any more for his thoughts on issues related to the submarine he had invented. A new skipper was appointed.

In April 1900, after twenty-five years of thinking about it, the United States Navy purchased the *Holland VI* for $150,000. Later in the year, six more Holland submarines in what became known as the Adder class were added to the contract. Elihu Frost and Isaac Rice, in the flush of success and enjoying substantial returns on their investment, were dismissive of the man who had started it all. They impugned Holland's intelligence and suggested that he should be content with his day in the sun. It was time, they said, for a new guard, one better qualified to guide the submarine project.

In 1904, John Holland formally resigned from the Holland Torpedo Boat Company. Although he made some attempts to put forward his design for a larger and faster submarine, contracts never materialised. John Holland, who had never earned more than $90 a week, enjoyed no personal fortune. He held one-half of 1% of stock in his company, if that. Any attempt he made to start a new company was blocked in court by the Electric Boat Company, claiming that Holland had assigned all rights and patents to them for his lifetime and had agreed verbally not to compete. Holland argued unsuccessfully that this was not so, but Rice and Frost were strong and determined opponents. John Holland found himself isolated and alone, his dreams for a bigger and better submarine doomed.

Holland, near-sighted and suffering from rheumatism, retreated to the workshop behind his home in New Jersey. In March 1914, less than a year after the death of his nineteen-year-old daughter, John Holland died at the age of seventy-three.

'He was a fair fighter, a most interesting and amusing companion, the staunchest of friends,' said Rear Admiral Kimball of his friend. 'God rest his soul.'

Internal combustion

RUDOLF DIESEL

Born: March 18, 1858, Paris, France
Died: September 30, 1913, somewhere in the English Channel

The golden hour of invention must terminate like other
hours, and when the man of genius returns to the cares,
the duties, the vexations and the amusements of life,
his companions behold him as one of themselves,
a creature of habits and infirmities.

Isaac Disraeli

THE DOOR TO THE SMALL PARIS FLAT CLICKED SHUT as Theodor Diesel, a German leatherworker, his wife Elise and their two daughters left for an afternoon in the park. Eleven-year-old Rudolf sat alone in the shadows, tied to a chair, his punishment for taking apart the family's cuckoo clock. A few months later, Rudolf, caught fibbing, walked to school wearing a sign that read 'I am a liar'. In spite of his austere upbringing, Rudolf grew up to be handsome, sociable, shy and soft-spoken, the master of three languages and a world-famous inventor. He was also plagued by personal demons, enduring chronic headaches and ill health and harbouring a crippling fear of poverty.

Life changed for Rudolf Diesel when France declared war on Prussia in July 1870 and all foreigners were ordered to leave Paris. The Diesel family, virtually penniless, fled to England, while twelve-year-old Rudolf was sent to live with an aunt and uncle in Augsburg. At last, he had a happy home and access to a fine technical education.

Rudolf Diesel's early inventions involved refrigeration and resulted in his first patent, for a machine that produced crystal-clear ice. When, under French law, all profits from the ice-making machine reverted to his employer, Diesel turned instead to the development of the new and highly efficient engine that he had in mind. In 1892, Diesel applied for and obtained a development patent for a combustion engine that was much smaller and more efficient than any engine then being produced. Diesel was confident that his engine could be developed quickly, but, in the end, the intricate problems around injecting and atomising fuel would take years to resolve. With no working capital of his own, Diesel sold the German rights to Maschinenfabrik Augsburg in 1893 and the patent rights for Austria-Hungary to the Krupp Werke in Essen. The Sulzer Company bought rights to the Swiss patents. The two most heavily invested firms, Augsburg and Krupp, pooled their formidable resources to produce a prototype.

Early experiments showed promise, but the engines were not yet truly successful. Over five years of revised drawings, redesigned parts, test runs and bitter disappointments, Diesel suffered from increasingly severe headaches and deteriorating health. Desperation crept in as he experimented with different fuels and struggled to perfect the fuel injection system, but the two manufacturing partners never wavered in their support. In 1896, the first of a new series of models was ready, burning low-grade inexpensive kerosene to produce double the power of existing engines. Krupp considered withdrawing its backing in the face of a number of patent infringement suits, but Diesel was convinced that the claims were spurious and the manufacturer stood firm.

By his fortieth birthday in 1898, Diesel was a millionaire many times over from the sale of rights to his engine. Among those interested in the Diesel engine was Adolphus Busch, who had made a fortune producing beer in America. Encouraged by glowing reports, Busch moved to establish himself as the sole representative for the Diesel engine in the United States and a visit to Germany presented just the opportunity to close the deal. In September 1897, Busch sailed across the Atlantic, trailed by a retinue of more than fifty people, including relatives, friends and staff. Their destination was the popular spa, Baden-Baden, where Busch and his family would take the waters.

Arriving at Baden-Baden in early October, Diesel found Busch ensconced in a suite that occupied an entire floor of one of Germany's most luxurious hotels. Busch, his pockets literally bulging with gold pieces, proved to be pleasant and unpretentious as the two men talked about an exclusive American and Canadian franchise for the engine. When Diesel asked for a million marks, Busch didn't hesitate before writing a cheque. On January 1, 1898, Adolphus Busch's Diesel Motor Company opened its offices at 11 Broadway in New York.

Augsburg-Krupp held the basic manufacturing rights to the Diesel engine in Germany, but Diesel himself held the rights for most other countries. Visitors interested in a superior combustion engine continued to stream through the factory. Among them was Alfred Nobel's brother, Emmanuel, who obtained the rights to manufacture and distribute the Diesel engine in the Russias.

As the popularity of the Diesel engine spread and negotiations became too complex for one man to handle, a group of wealthy backers formed the General Diesel Company in 1898 and purchased all rights and patents from the inventor for 3.5 million marks. Two thirds of the selling price was accorded in the form of stock. Within a short time, Diesel was to see the same amount of money that he had received for all rights to the engine change hands in one single transaction for partial rights under the leadership of the General Diesel Company. Rudolf Diesel had sold himself short.

As rumours began to circulate that the Diesel engine wasn't all it was professed to be, Diesel's income increasingly fell short of his growing expenses. In the late 1890s, Rudolf and his wife, Martha, began construction of a mansion in Munich, complete with an indoor bicycle track for the children to use on rainy days. Five bathrooms were equipped with the most up-to-date equipment; staff included a butler, a governess and other servants. Martha, very much involved with Munich's social set and anxious to keep up appearances, was increasingly distant from her husband and his 'black mistress'. The gentle and highly anxious inventor, besieged by almost constant excruciating headaches, wrote that he and Martha were 'no longer spending time together'.

Just as the family prepared to move into their mansion, news came that the Balkan petroleum venture in which Diesel had invested was in trouble. Manufacturing and performance issues continued to plague the engine, and stock in the General Diesel Company plummeted. Persuaded to purchase even more stock in order to restore confidence in the engine, Rudolf suffered mammoth losses when the company eventually went bankrupt. That loss, coupled with the stunning bills for his new home, decimated Diesel's financial reserves.

In the late 1890s, Diesel invested in two new inventions, the telephone and the typewriter. He may well have been sorry: soon the telephone started to ring with bad news about stock prices, licencing disagreements and difficulties with the engine. Suffering from nervous exhaustion, Diesel was unable to sleep. The luxurious mansion was, in his words, 'a mausoleum', but when he suggested they sell it, Martha refused. Feeling out of place in his luxurious new study, Diesel soon transformed it into a replica of his old Augsburg workshop. Convinced that his personal destiny was wrapped up with the future of his engine, an invention that would ultimately benefit all mankind and improve the lot of the working man, Diesel converted the top floor of his mansion into a construction office, furnished with long drawing tables and peopled with engineers and draftsmen. Visionary drawings of engines for use in boats and ships, aircraft and cars papered the walls but, as Diesel himself bemoaned, 'it all took so much time'. Developments were not moving fast enough to keep up with his financial needs. Rudolf had made and lost a fortune and time was running out. He had been, he claimed, born too early for the work he wanted to do.

The Paris Exposition of 1900 awarded its highest honour, the Grand Prix, to the Diesel engine, but by 1907 Diesel was nearly half a million marks in the red. His financial position continued to deteriorate until, in 1912, he had suffered losses amounting to more than ten million marks. Servants were discharged as Diesel, a pacifist, became increasingly depressed about the world situation.

In 1912, Rudolf and Martha sailed for New York, encouraged by Adolphus Busch, whose agency for the Diesel engine was floundering. When Busch changed the date for the groundbreaking of his new Diesel factory, the Diesels accommodated and sailed earlier. They were to have travelled on board the *Titanic*.

To boost interest in his invention in America, Diesel gave speeches and interviews. He foresaw the introduction of his engine into trains, as it had already been introduced into sailing ships. With great vision, Diesel predicted that air pollution would become an important consideration in the future. On the eve of the automobile age, the Diesels' reception was enthusiastic, enhanced by Thomas Edison's characterisation of the new engine as one of the 'truly memorable accomplishments of mankind'.

In May 1913, at the end of their successful visit to America, Martha and Rudolf took the train to Orange, New Jersey to visit the king of inventors, Thomas Edison. Edison, dressed in work clothes, greeted the Diesels in his cluttered workroom. A cot stood in the corner and his famous wooden roll-top desk overflowed with papers. The only armchair was made of cement. As the hours passed, Diesel realised that this was not the magical meeting of minds that he had envisioned. Edison, who emphasised proudly that he was self-educated, knew a little about a lot of things, including the Diesel engine. Rudolf Diesel, whose knowledge of things electrical was limited, recognised that Edison was uneducated in the two fields considered key for an inventor, mathematics and physics. Edison did not endear himself to the German inventor when he expanded on his intention to bring together ideas from ten or more inventors to create the perfect engine. Diesel acerbically enquired as to whether Mr Edison thought engine inventors were worth but a dime a dozen. Collaborative by nature, Diesel actively disagreed with Edison's assertion that the truly successful inventor was no more than a creative coordinator of other men's ideas.

The visit, scheduled to last a week, ended after five days. Edison fell in Diesel's estimation and Edison, who tended to pooh-pooh anything artistic and cultured, found Diesel to be 'Bohemian' and 'eccentric'. Meals generally consisited of vegetables; Edison, who was teetotal, never served alcohol. Diesel was heard to remark that Edison was truly 'a glutton for the simple life'. As the Diesels drove away, Edison called after them, 'Don't eat so much'. Thus ended the meeting of two of the greatest inventive minds of the century.

In 1913, Martha left Munich to visit her mother. Diesel, at home in their mansion, gave the servants a long weekend and put his younger son on a train for Switzerland. Alone, Diesel asked his older son to spend a few days with him and, while they were together, Rudolf shared his personal papers. In Frankfurt, Diesel joined his wife and his daughter's family, including a new grandchild. On parting, he gave his wife an elegant travelling case, instructing her not to open it for a week. His reading matter on the train to Ghent was Schopenhauer, the bookmark left in the section on wealth and its management.

Rudolf Diesel joined friends for the Channel crossing to England. After dinner on the boat, they parted company to return to their cabins. Diesel left a wake-up call for 6.15 on the morning of September 30, 1913. When he didn't appear for breakfast, his friends went to his stateroom to find his bed untouched and his nightshirt laid out. Diesel's hat and overcoat were found, folded, by the deck railing. On reaching port, the esteemed inventor Dr R. Diesel was reported missing.

When his wife opened her case a week later, she found all the cash that her husband had been able to pull together. Ironically, the real estate investments that initially spelled his ruination increased enormously in value after the First World War.

Twenty-three years after Rudolf Diesel's death, the Mercedes-Benz 260D became the first mass-produced passenger car to be powered by a Diesel engine.

George Selden is pictured on the right

Road hog

GEORGE SELDEN

Born: September 14, 1846, Clarkson, New York
Died: January 17, 1922, Rochester, New York

I have the result, but I do not yet know how to get to it.

Karl Friedrich Gauss, German mathematician

GEORGE SELDEN, AN INVENTIVE MAN IF NOT AN INVENTOR, was a visionary of a different stripe. A patent attorney by profession, Selden saw the potential of the car early and, although he didn't intend to manufacture such a vehicle in earnest, he described in a patent what he saw as the prototype for the future — a basic road vehicle propelled by an internal combustion petrol engine.

In 1878, Selden teamed up with two machinists in a small Rochester, New York workshop to create a compact engine, a third the weight of an earlier two-stroke prototype. Although Selden and Gomm couldn't get their engine to run for more than five minutes at a time, Selden felt it was good enough to warrant a patent for his 'Road Engine' and its use in a four wheeled car, for which he applied in 1879. Under the law, Selden was required to submit a working model, but he was allowed, instead, to describe the general features until some future date. Here, Selden's expertise as a patent lawyer served him well. At the time, an inventor was allowed two years to complete a patent application; Selden managed to extend that to nearly seventeen years by filing a hundred amendments incorporating technical improvements. The final patent for his engine was issued in 1895, and provided an air-tight patent roadblock to serious manufacturers.

As far as Selden was concerned, it wasn't the engine that mattered, it was the idea.

The important fact was that he was the first in America to apply for a patent covering petrol-powered cars, and from 1895 onwards, Selden was in the enviable position of extracting royalties from American manufacturers for every car they built. In November 1899, the Electric Vehicle Company of Hartford, Connecticut, headed by William Whitney and Colonel Albert Pope, paid Selden $10,000 and royalties for his patent and began to extract royalties from other manufacturers, just as Selden had done.

By 1903 ten leading motor manufacturers – including Cadillac, Packard, Locomobile and Peerless – had joined together to form ALAM, the Association of Licensed Automobile Manufacturers, to control the manufacture of cars. Advertisements were posted threatening manufacturers and purchasers with prosecution if they sold or used a car which had not been licenced under Selden's patent. Brass plaques were affixed to all ALAM cars bearing Selden's patent number, and ALAM spies canvassed the streets checking parked vehicles for the proper validation.

A few weeks after ALAM was formed, the Ford Motor Company was incorporated and Henry Ford's application for a manufacturing licence was rejected out of hand by ALAM. Refusing to be intimidated, Ford continued to sell cars. His stockholders nervously voted to stand their ground and defy the Association. In October 1903, ALAM filed suit against Ford.

In 1909, six years after ALAM sued the Ford Motor Company, the monopoly case went to court. At one point, the trial was halted while everyone went to the window to watch the start of the Transcontinental Auto Race.

'There is something that puzzles me,' Ford said to the judge. 'I see a Ford car, two Ford cars, but I see no Selden' – referring to the fact that no working Selden prototypes existed.

The judge appreciated the joke, but ruled, on September 15, 1909 for the monopoly. 'While George Selden did not invent the different components that made up his car,' the judge stated, 'he was the inventor of the ensemble.'

In the wake of the legal finding, other manufacturers who had stood firm with Ford caved in and begged for membership of ALAM. Ford was now completely isolated, but his ire was raised; when ALAM offered concessions, he refused. In 1909, under increasing pressure, Ford considered selling his company to William Durant, who was in the process of forming General Motors, but when Ford demanded the $8 million selling price in cash, Durant baulked. His bankers were not convinced that Ford was worth anything like that sum.

In 1910, Ford appealed the 1909 verdict. In preparation for the trial, Selden had been instructed to build a car to his patent, with pitiful results. Ford's lawyer, Fred Coudert, ridiculed the Selden car, claiming that it couldn't do better than one thousand, three hundred and nine feet in an hour and twenty minutes.

Summoned to a pre-trial meeting with the ALAM counsel, Coudert was asked to take a seat in the waiting room. To pass the time, he picked up a new book by Scotsman Donald Clerk, who had testified as an expert witness for ALAM at the first trial. Coudert was astounded to read that Clerk had now changed his mind. Revising his opinion, Clerk advocated that Selden was not entitled to his pioneer patent on the grounds that it reflected advances carried out earlier by Otto, Daimler and Benz. Clerk was publicly contradicting the testimony he had given at the first trial. On January 9, 1911, the Appeals verdict came down strongly in favour of Ford.

Henry Ford, hailed as a hero, received letters applauding his courage from friends and foes alike. When he arrived at the ALAM banquet, Ford was greeted with cheers, in response to which he bowed but said nothing. The publicity accorded to Ford through the trial was some of the best advertising he could have wished for his new Model T.

In November 1912, George Selden's patent for his Road Engine expired once and for all. In that same year, the Supreme Court revised the patent laws that Selden had so cleverly exploited. For a decade, George Selden had held the automobile industry to ransom with a patent for a car that didn't work. In the process, he had accrued hundreds of thousands of dollars in royalties for doing virtually nothing.

George Selden suffered a stroke in 1921 and died on January 17, 1922.

Eye catcher

PERCY SHAW

Born: 1890, Yorkshire, England
Died: 1976, Yorkshire, England

*The greatest invention of the nineteenth century
was the invention of the method of invention.*
Alfred North Whitehead

HOME FOR MR AND MRS SHAW AND THEIR FIFTEEN CHILDREN was a small eighteenth-century house without gas or electricity on a hillside in the village of Boothtown, near Halifax, in Yorkshire. Every one of the Shaws either sang or played an instrument – Percy played the flute and the violin – and any free time was spent making music.

When Percy left school at the age of thirteen, not an uncommon occurrence in those days, he found work at a blanket factory. In the evenings, he studied shorthand and bookkeeping in an effort to move on to a better job. Tiring of his studies, Percy tried his hand at a number of trades, in the course of which he devised a process for backing carpets with rubber.

When Mr Shaw, who earned a pound a week as a dyer's labourer, lost his job, Percy joined his father in setting up an odd-job company that operated out of their barn. The First World War offered just the opportunity father and son had been waiting for, and a contract was awarded in 1914 for the manufacture of special wire used in making khaki leggings for the soldiers. The Shaws also invested in a forge and anvil for making shell noses and cartridge cases.

Most of the Shaw children married and moved away, but Percy stayed put. By

1930, with both parents gone, Percy was 'Master of the House'; he was now 'Mr Shaw'. The barn workshop buzzed as he developed his new project, the building of private roads and garden paths. Business was good and Percy hired a few workers.

Electric trams were a popular means of public transport then, running on raised steel rails embedded in the road. The stretch between Queensbury and Boothtown was particularly dark and treacherous, with one side falling off in a steep drop. As Percy drove 'Doris' – the Model T Ford he had bought in 1916 – home along that road through the fog, he noticed that the tram lines shone through 'like twin silver ribbons'.

One night, on a particularly dark stretch, Percy Shaw lost his bearings. 'I crawled along,' he said, 'feeling sick at the thought of the death drop down the side, with nothing more than a frail fence as protection.' Then, according to his story, two points of light pierced the darkness – a cat, sitting on a fence – and those points of light saved him from almost certain death. The cat's eyes and tram tracks gave birth to a new idea and a fortune in the making.

With a plan for his new invention, a reflective road stud, Percy, who had never gone further afield than London, now decided to journey to Czechoslovakia in search of pure crystal glass suitable for the studs he had in mind. Joe Horton, his brother-in-law, went with him. The small factory in the barn that had housed all the Shaw enterprises now hummed with Percy's new project. A small furnace was installed and a testing tunnel was fashioned from old oil cans.

The time Percy Shaw had spent as a jack-of-all-trades served him in good stead. With a prototype of the stud in hand, Percy left his factory in the dark of night to try his new invention on the road. Under the guise of 'road repairs' he dug up a small portion of tarmac and inserted a stud. A few hours later, he pulled up the stud and went home, a process he repeated many times. Perhaps the most ingenious aspect was his method of keeping the stud clean – using a rubber cushion to collect water and close over the stud, as an eyelid closes over an eye, when a tyre ran over it.

A man of few means and with little worldly experience, the butt of a great many jokes as he persisted in his work, Percy Shaw understood that he had a unique product, and in April 1934 he wisely took the precaution of protecting his invention with a patent for 'Catseyes'. There was competition, but Shaw's product was superior. In America, the Road Bug, a version of catseyes, died a death because of poor design, and a Mr Walton from Blackpool devised a reflector but failed as a salesman.

The powers that be in Britain agreed to entertain a demonstration of catseyes in 1935 – at the inventor's expense – and the experiment went well, but a full year passed before the first order came in for thirty-six studs for a pedestrian crossing near Bradford. Orders for a hundred here and a hundred there trickled in. In March 1935, Percy Shaw registered his trademark and formed a company with two co-directors. In various tests, the catseye was the only reflector that performed satisfactorily and Percy's invention received official recognition, but it wasn't until the Second World War that Percy came into his own. 'Fifty thousand catseyes help fight the blackout', screamed the headlines and orders poured in, including a standing order for forty thousand studs a week.

The factory was expanded, but remained on the site of the original barn where Percy had worked his entire life. His favourite sycamore tree, which he refused to sacrifice, was accommodated and grew up through the new building. In 1947, Percy Shaw was chosen as one of the most notable northern personalities, appearing on the radio with Richard Dimbleby in *Down Your Way*. The House of Commons described the catseye as 'the most brilliant invention ever produced in the interests of road safety'.

In July 1949, Princess Elizabeth visited Halifax for the town's centenary. When she saw catseyes exhibited in a trade exhibition, she was heard to remark, 'Oh, I think they are the most wonderful things on the road,' and asked to meet the inventor.

A millionaire many times over, Percy Shaw, a confirmed bachelor, never left his family home, and made few concessions to modern conveniences. The only luxury he afforded himself were two Rolls Royces, which he proudly parked in the garden beside his house on the hill.

Cooking caoutchouc

CHARLES GOODYEAR

Born: December 29, 1800, New Haven, Connecticut
Died: July 1, 1860, New York City

Life should not be estimated exclusively by the standard of dollars and cents. I am not disposed to complain that I have planted and others have gathered the fruits. A man has cause for regret only when he sows and no one reaps.

Charles Goodyear

The objective for Charles Goodyear was little less than a religion.

James Wickham Roe, Yale University

CHARLES GOODYEAR WAS A MAN OBSESSED, a trait shared by many other inventors, if not always to this degree. Money and fame were the least of it. Goodyear was, quite simply, entranced by... rubber!

While Goodyear migrated to and from debtors' prison following the failure of his hardware business, Englishman Thomas Hancock opened the world's first rubber factory across the Atlantic. Glasgow chemist Charles MacIntosh had tempered rubber with ammonia to produce waterproof garments comprising two coated layers enclosing a core of gum. Rubber was used in Britain with some success for hoses, surgical instruments, shoes and pipes in the breweries. The promise of crude rubber had been recognised, but to date no one had succeeded in adapting it for practical daily use.

Caoutchouc, or gum elastic, is believed to have been brought to Europe from South America by Columbus. For three hundred years, no use was found for the substance until English inventor Joseph Priestley discovered that gum was effective for rubbing out pencil marks and produced the eraser. Rubber made its appearance in America in 1823 with the importation of gum shoes from Brazil. Imagining the fortunes to be made in this new substance, millions of dollars were poured into the rubber business in New England.

The problem was that New England's wide range of temperatures soon revealed inherent flaws in this miraculous new substance. In very cold weather rubber became brittle; in the heat of summer it was distinctly sticky and smelly. Customers wearing rubber garments were warned not to stand too close to the fire. Shoes hardened, braces sagged, raincoats rattled like tin cans or stuck together in unwieldy folds. Factories were abandoned, fortunes lost, and great bonfires of useless rubber articles set alight.

One of Charles Goodyear's peculiarities — an anomaly in a person who had no experience with the sea — was his lifelong fascination with aquatic life-saving equipment. Out for a walk in New York one day, Goodyear was attracted by a display of rubber life preservers in the window of the Roxbury India Rubber Company. He entered the store and bought one of their life rings. When Goodyear returned a few days later with his ingenious design for a new valve, the owner told him about the disastrous turn in the rubber industry, not least the $30,000-worth of shoes that had been returned because they stuck together and emitted a terrible odour. The company, he confided, would pay any price for the solution to rubber's eccentricities.

This, Goodyear was convinced, represented his fortune and his life's mission. Rubber was the field of invention for which he had been searching. Now, under house arrest for debt and with no working knowledge of chemistry, Goodyear undertook his first experiments using the kitchen in his family's small cottage as his laboratory. Neighbours complained about the terrible smell at the Goodyears' house, so one can only imagine what it was like for the immediate family. During one experiment, Goodyear's assistant dipped his trousers into a barrel of uncoagulated latex, only to find himself, a few minutes later, firmly cemented to his seat, his legs glued tightly together.

Goodyear's early experiments with rubber drew the family further into debt. It is said that when a stranger asked how he might recognize Charles Goodyear, he was told to look for a man wearing an India Rubber cap, vest, coat and shoes and carrying an India Rubber money purse with not a cent in it. At one point, Goodyear was so involved in his new acid gas process that he was overcome and had to be dragged from his small workroom.

Some backers came forward, but the depression of 1836 and the earlier collapse of the rubber industry frightened off potential investors. Goodyear and his family were reduced to penury. Living in a small cottage on Staten Island, New York, he was once forced to leave his umbrella with the ferryman in lieu of fare. Whenever Goodyear did find himself with a little extra money, the funds went directly back into his experiments. Urged to return to the hardware business after four years of failed attempts, Goodyear exhibited the same single-mindedness and sense of divine purpose exhibited by many other great inventors. He was, in his words, driven, 'lest the discovery should be lost to the world and himself'.

Nathaniel Hayward, a partner in Boston's Eagle India Rubber Company, had developed a process combining all the key ingredients of vulcanisation. It was Hayward who introduced Goodyear to the magic component, sulphur. Exposure of rubber to a consistently high degree of heat, the last step in the vulcanisation process, needed to be refined, but they were close to a solution. Goodyear encouraged Hayward to patent his process, which he did in 1838. Hayward immediately assigned his patent to Charles Goodyear in exchange for one hundred dollars, with nine hundred more to be paid in six months. Hayward retained the right of manufacture until the last payment of two thousand dollars had been made. Tragically before that time came rubber goods produced at the factory proved unsatisfactory. Goodyear let Hayward go and was returned to a state of penury.

Legend has it that the next step was accidental. A sample of Goodyear's sulphur-treated rubber came into contact with a very hot stove and, rather than melting, the rubber charred. The key, Goodyear saw, was to interrupt the charring process at just the right point, a revelation that ushered in the five bitterest years of his struggle. Suffering from chronic dyspepsia, he worked alternately from his jail cell or his sickbed, where he spread his experiments out on his counterpane. He sold anything he could find, including his children's schoolbooks, to finance his work. It was a common sight to see the Goodyear children in a state of extreme destitution, digging for potatoes in the fields behind their cottage.

Searching for the perfect heat source, Goodyear boiled his specimens in saucepans, held them in the tea kettle and roasted them over the fire. He begged the use of large boilers and experimented with open brushwood fires. The gum either blistered, melted or charred, but Goodyear persisted. Over the course of 1840 and 1841, Goodyear virtually lived in his drab rubber raincoat, covered with dark spots where he had heat-tested the material.

In December 1841, Goodyear had a Justice of the Peace draw up specifications for the curing process – pure gum compounded with sulphur and white lead, to which heat was then applied. These specifications were filed as a caveat with the Patent Office, affording the inventor one year to perfect his process. Unfortunately, his patent for the vulcanisation process, issued in June 1844, left fatal loopholes. In July 1844, Goodyear sold the valuable rights to manufacture under his patent for fifty thousand dollars to a company that later became Uniroyal.

In 1841, Goodyear entrusted rubber samples to an Englishman, Stephen Moulton. The MacIntosh Company declined to spend the $50,000 Goodyear was asking for rights to his process, but Englishman Thomas Hancock, a partner in the MacIntosh company, was quick to realise the significance of Goodyear's discovery. What he saw provided the clues he needed to perfect his own process and in November 1843, Hancock applied for a provisional English patent, a few weeks before Goodyear filed in the same country. Goodyear's application was denied. Under oath in court, Hancock admitted that 'the first vulcanised rubber he ever saw came from America'. Goodyear never realised profits from his discovery in England or France, having been refused a French patent on the grounds that he had sold vulcanised rubber products commericially, albeit in small quantities, in France prior to his patent application.

In the words of his son William, Charles Goodyear both 'created and despised wealth'. Frugal in every other regard, Goodyear spent lavishly on his experiments. He conceived, developed and financed idea after idea, then sold out for a bonus and royalties, usually far below market value. Charles Goodyear had a certain contempt for money except as a means to an end. Ever the inventor, he could not make the move to manufacturer. It didn't help that over time the legal defence of his patent required far more money than Goodyear had to invest.

Within seven years of filing his vulcanisation patent, Charles Goodyear found himself in the middle of one of the greatest American business law suits of the nineteenth century, the Great India Rubber Case. The decision, handed down in September 1852, represented a total vindication of Goodyear as the true inventor of vulcanised rubber. The reissue of his patent was upheld.

In 1853, Goodyear published a 620-page book on the possible uses of gum elastic. Some copies were printed on rubber instead of paper. The treatise, a testament to Goodyear's rich imagination, included rubber sails, a watch with a rubber case, cutlery with rubber handles, picture frames and tools, all fashioned out of rubber. Goodyear proposed an embossed rubber globe that could double as a football when not in use in the classroom.

Not least among Goodyear's ideas was the rubber band, 'for use by druggists and tradesmen for tying up small packages'. He created a water bed, a box filled with water and covered over with a rubberised cloth, 'comfortable for the ill and lame'. Sports equipment included the inflatable ice skating cap and inflatable boxing gloves. He proposed rubber musical instruments, washable rubber bank notes, money belts (also adaptable as life preservers) and rubber wheelbarrow tyres. A chapter in the book entitled 'Life Preserving Travelling Apparatus' listed items such as trunks, valises and hat boxes that could be lashed together to form rafts.

When Queen Victoria opened England's Great Exhibition on May 1, 1851, Goodyear was ready with an exhibit calculated to knock the collective socks off the British. His 'Vulcanite Court', a suite of three rooms furnished with all things rubber, cost $30,000 and occupied more space than all the other American exhibits combined.

During the 1850s, Goodyear was ill much of the time. His overall financial picture improved, although he was still heavily in debt. In 1855, Goodyear mounted an exhibition at the Exposition Universelle in Paris, taking out $50,000-worth of loans that, ultimately, he was unable to repay. On December 5, 1855, Goodyear was asleep in a Paris hotel with his wife when two officers knocked at the door and the gendarmarie climbed in through the window. Goodyear was removed once again to debtors' prison. Louis Napoleon, who had visited Goodyear's exhibit and was highly impressed, made Goodyear a Chevalier de la Légion d'Honneur. The medals were delivered to the prison by Goodyear's son Charles. When Goodyear was released, he returned to find that his hotel room had been redecorated in his honour as a Chevalier de la Légion d'Honneur.

After a difficult trial in 1858, Goodyear's US patent was extended. In his last year of life, he moved into a house in Washington, DC in which he installed a large tank for the testing of life-saving apparatus. His death on July 1, 1860 was barely noted. He merited two paragraphs in the *New York Times*, and received no mention in the London *Times*. At the time of his death, Goodyear's affairs were in confusion. His attorney calculated that his receipts from every source since 1835 amounted to $368,880.23; the claims on his estate came to $347,983.67, leaving a total of not quite $19,000 to show for twenty-nine years of struggle and deprivation.

Goodyear's son Charles, who had lived much of his life in dire poverty, interested himself in the shoe business and became a wealthy man through the promotion of others' inventions. In 1865, the year the vulcanisation patent expired, Charles Junior sold the Goodyear name to a Californian company.

The Goodyear patent for the vulcanisation of rubber expired on June 15, 1865. During his lifetime, many people believed Goodyear to be an extremely well-to-do man; after all, his name appeared on the patent label on dozens of everyday items. That misconception led to misplaced envy and antipathy. Many today believe that the inventor is connected with the multinational corporation that bears his name; in fact, Goodyear Tyre and Rubber, established in 1898, did nothing more than borrow from history.

Painted by Mc King. Wm Hoogland Sculp.^t N.Y.

Eli Whitney

Running wild

ELI WHITNEY

Born: December 8, 1765, Westborough, Massachusetts
Died: January 8, 1825, New Haven, Connecticut

An invention can be so valuable
as to be worthless to its inventor.

Joseph Wickham Roe, Yale University

ON APRIL 30, 1789, THE DAY GEORGE WASHINGTON WAS INAUGURATED
as the first President of the United States, twenty-three-year-old Eli Whitney, a farm
boy from Westborough, Massachusetts, walked onto the Yale campus to begin his
education. Eli was not brilliant academically, but he was an eager student.

On April 19, 1775, the British forces marched on Lexington and Concord and
the 'shot heard 'round the world', marking the beginning of the American Revolution,
landed about twelve miles from the Whitney farm. With the advent of war, Eli
persuaded his father to install a forge for the manufacture of nails and, using crude
homemade machinery, father and son built a thriving business. In 1783, with
inexpensive goods once again flowing into the country, Eli Whitney turned his hand
to making steel hatpins for ladies.

Whitney went on to teach for five years in order to earn the money necessary to
study law at Yale. Armed with his degree and heavy debts, Whitney accepted a tutoring
position in Georgia. Penniless, he set sail for Savannah in 1793. Among his
companions on the trip were Phineas Miller, a lawyer and Yale graduate, and Mrs
Nathanael Greene, the widow of a Revolutionary War general. Mr Miller and Mrs

Greene issued an invitation to Whitney to visit Mulberry Grove, Mrs Greene's home in the South.

One afternoon there, seated with his new friends in the parlour, Eli Whitney joined in conversation with local planters. Because it took a skilled worker a full day to clean the seed from a pound of cotton, some new method was urgently required to make the crop economically viable. Mrs Greene suggested to Whitney that he study the problem.

With a sack of raw cotton, tools he had crafted himself and wire intended for use in a birdcage, Whitney retired to a room in the basement. Within ten days he had produced a prototype of the cotton gin, the word 'gin' being an abbreviation of the word engine. There is ongoing controversy about the part that Mrs Greene played in the invention of the cotton gin, with many claiming that she provided the ideas. The truth of it has never been established for certain, but we do know that she provided the funding essential to the establishment of Miller and Whitney's new company.

In a letter to his father dated September 11, 1793, Whitney wrote that he had constructed a model and that someone had offered him 'a hundred guineas' for all rights and title. He declined. He also reported that it was generally said, by those who know anything about it, 'that I shall make a Fortune by my invention'.

'I have no expectation,' he continued in his letter, 'that I shall make an independent fortune, but I think I had better pursue it than any other business into which I can enter.' Eli acknowledged that something unforeseen might frustrate his expectations but, he added, 'I am so sure of success that ten thousand dollars, if I saw the money counted out to me, would not tempt me to give up my right or relinquish my object.' He swore his father to secrecy.

The first model of the cotton gin, which contained all the elements of the final machine and was very simple in design, virtually transformed America's cotton industry overnight. With the new machine, one person turning the crank could do the work of fifty people picking by hand. Before the winter of 1792–93, America had exported about a hundred and forty thousand pounds of cotton; in 1800, America exported nearly eighteen million pounds.

Whitney didn't need to convince anyone that his invention was worthwhile. In fact, the frenzy for access to his machine soon outran the marketing plan. Phineas Miller and Eli Whitney quickly formed a partnership, believing fully that they had found their way to a fortune, but they made two critical mistakes. One was their failure to shroud the new invention in secrecy until such time as the rights were adequately protected and the means of manufacture in place. Visitors to Mulberry Grove were invited to view the model, so simple in design that it was an easy matter to reproduce Whitney's invention from memory.

Even as Eli Whitney travelled to Philadelphia to obtain a fourteen-year patent signed by President George Washington on March 14, 1794, Miller entered a notice in the *Georgia Gazette* announcing optimistically the availability of sufficient numbers of gins to process the next cotton crop. The pressure was on. Whitney, working in his factory in New Haven, expressed valid concern about keeping up with sales. All hope was gone when the factory burned to the ground in March 1795, taking with it Whitney's tools and more than twenty completed machines.

Miller and Whitney's second critical mistake was their decision to own and operate the gins themselves, charging customers one-third of their profit. Selling the machines or the rights to manufacture the machines would have been both easier and more lucrative. In the way that Xerox leased their early machines and charged per copy or American Bell Company decided to rent out their telephones, every new idea requires a new approach to marketing. In Whitney's case, he saw the gap in the market but the market in the gap devoured the invention. When Whitney and Miller failed to supply enough cotton gins to meet demand, the invention became a juggernaut.

The crops wouldn't wait and before long copycat models of the gin appeared throughout the south. Whitney and Miller, who had lost control of their invention, argued for redress before state legislatures and in some instances were awarded compensation, but only enough to cover costs. On one occasion, defendants claimed that the cotton gin had never been used in Georgia, at the same time that three machines clattered away within fifty yards of the courthouse. The potential economic impact of the cotton gin negated all attempts to control infringement. A similar potential pertained when Daguerre first introduced the prototype of his camera and people responded with a passion for taking pictures. The French government, understanding that the invention was too dynamic to control under the patent system, bought the rights from the inventor and gave the camera to the world.

Litigation of nearly sixty patent infringement suits was long and costly, although Whitney's rights were upheld in 1806 when the judge declared one version to be simply a variation on Whitney's 'undeniably original principle'. Application for an extension of his original patent, due to expire in 1807, was refused. In total, Whitney received about $90,000 for his invention and spent nearly that amount in litigation costs. As a result of Whitney's painful experience with patent infringement, inventors Robert Fulton and Oliver Evans approached him later for help in modifying the patent laws.

Whitney was lucky to more or less break even. One lesson he had learned was that to make money through invention you needed to have money to promote and protect your idea and to manufacture your product. The idea alone was not enough.

Eli Whitney died a rich man, but that wealth had only indirectly to do with the cotton gin. It was a contract from the government to manufacture firearms that led to his fortune. Nearly bankrupt, no longer enjoying the support of Catherine Greene or Phineas Miller, and accepting that an ongoing fight for his rights in regard to the cotton gin would prove fruitless, Whitney wisely turned his attention elsewhere, even before the controversy had been resolved.

Only six years had passed since the invention of the cotton gin and Whitney was just over thirty years old when, in 1798, he signed a contract with the American government worth $134,000 for the manufacture of 10,000 firearms. What made this arrangement so remarkable was that Whitney had no experience, no factory, no employees, no machines and no stock. A man who had never before manufactured a gun was soliciting the largest order for firearms ever placed in America and he had just two years to complete the contract. As ill-equipped as Eli Whitney was to undertake such an ambitious contract, he had rescued one very important asset from his experience with the cotton gin, his reputation, and it was on this that Congress was banking.

The next ten years – for that is how long it ultimately took Whitney to complete the contract – were stressful, as deadlines came and went. Reluctantly, many times over, Whitney was granted additional months to complete his government contract. His home and factory were mortgaged; everything rested on successful completion of the contract. Whitney admitted to having nightmares about the scope of his commitment, but recognised that it was too late to turn back.

The situation was fruitful in unforeseen ways. Thomas Jefferson, visiting Europe in 1785, had witnessed a process for the manufacture of muskets using standardised parts and was impressed by what he saw. Whitney, who was establishing his plant in Connecticut for the manufacture of thousands of firearms, was quick to pick up the idea. Using water-powered machines, each performing one step in the fabrication of uniform parts, Eli Whitney went on to pioneer mass production using standardised, interchangeable units. His perfection of this production principle was to transform the manufacture of goods in America and stands as possibly his greatest innovative accomplishment.

Eli Whitney missed having a family of his own and wistfully signed himself in letters as a 'solitary old bachelor'. To compensate, he surrounded himself with nieces and nephews. In particular, he cared for the three sons of his widowed sister and put one of them, Eli Whitney Blake, through Yale. Blake went on to invent the stonecrusher that bore his name, a machine that formed the basis of the concrete industry.

Whitney married for the first time in 1817 at the age of fifty-one. He and his wife, Henrietta Edwards, who was twenty years his junior, had one son, Eli Whitney II. When prostate disease caused Whitney agonising pain he turned his inventive hand to devising medical instruments to alleviate his suffering. Eli Whitney died on January 8, 1825. In that year, the United States produced fifteen million pounds of cotton.

Whitney's nephews ran the armoury until Eli Whitney II had graduated from Princeton and was able to take over the management. Eli, the son, held numerous patents having to do with the manufacture of guns and is credited with creating the first public water supply in New Haven. Whitney's armoury produced Colt's pistols for many years.

Confidence man

WILLIAM THOMAS GREEN MORTON

Born: August 9, 1819, Charlton, Massachusetts
Died: July 15, 1868, New York City

*That so unlikely an outcome should accrue to a man
possessed of such limited talent and so many flaws,
and one lacking in a sense of ethics and decency was
one of the bitter ironies of history.*

William Morton

ONE OF THE MOST PROFOUND DEMARCATIONS IN WORLD HISTORY was
the common application of anaesthesia. If ever there was a crying out for innovation,
this was it. Prior to the mid-1800s, people routinely chose death over surgery without
pain relief. Those who braved the knife were so traumatised that the very meaning of
life was often permanently changed for them.

To adequately appreciate the hunger for pain relief, imagine yourself in 1840.
You have a tumour; surgery is a matter of life or death. Through the night, the terror
builds. In the morning, you walk to the surgeon's office where you are strapped into
a chair. A tray is set up next to you, where a chisel, a saw and knives of all descriptions
are laid out. Men come to hold you down. The surgeon picks up one of his knives and,
trapped, you cry for mercy. If you're lucky, the surgeon is skilful and you recover, not
only from the surgery but from the dark place you have been.

This is a tame description compared with those that have come down through history, but I wanted to give some feel for the urgent need for advances in surgical technique. Electric lighting was a luxury; anaesthesia was life's blood. But, as pressing as the need was, the majority of individuals in the mid-nineteenth century, including medical professionals, were unable to envision the advent of painless surgery. It was a blind corner. At the other end of the spectrum were those protecting the status quo in the belief that suffering, and in particular the suffering associated with childbirth, were man's – and woman's – just desserts.

The motto of this story is that success depends on what you're looking for. Enter five individuals, actors in this account of innovation and discovery, only one of whom, William Morton, is connected in the popular mind with the introduction of anaesthesia. The room where Morton first administered ether on October 16, 1846 is preserved as a museum at Boston's Massachusetts General Hospital. Morton's name appears on the commemorative plaque and it is he who is immortalised in the oil painting of that momentous occasion.

Why Morton? Why not any of the other five physicians who not only appreciated the properties of ether and nitrous oxide, but who published and talked about their findings? And why, in fact, didn't any of the thousands of people who enjoyed public demonstrations of nitrous oxide – commonly known as laughing gas – at fairs and carnivals draw the dotted line between the joke of harmless unconsciousness and relief from the crucible of physical pain? It stands as a good illustration of the inventor's dilemma: how and when to go public. It is a step that often requires a lot of courage and can go badly wrong, as Jonas Salk was to find out.

Humphry Davy was the first scientist who had the chance to stamp his name in perpetuity on the discovery of anaesthesia. A master chemist and precocious experimenter, Davy challenged existing hypotheses and invented all manner of new chemical apparatus while working for Sir Thomas Beddoes' Pneumatic Institute near Bristol in the early 1800s. Numerous other inventions are attributable to Davy, including the Davy lamp used by miners. His was an enquiring mind.

In his time with Beddoes, Davy took exception to the view that inhaling nitrous oxide was harmful. Quite the opposite; he stated contrarily that nitrous oxide produced euphoria, mirth and, eventually, unconsciousness. In his 1800 volume on nitrous oxide, Davy drew a direct line to anaesthesia, stating that the gas 'could probably be used to advantage during surgical operations'. Ironically, Davy turned his attention to experiments with electro-chemistry, referring to his work with nitrous oxide as 'trivial' and the stuff of 'youthful dreams'. To some degree, Davy's creative vision was blurred by a belief that his studies with nitrous oxide, ultimately of incredible significance to humankind, didn't have the scientific cachet of other investigations.

The second figure in our story, Henry Hill Hickman, practiced as a country doctor in Shropshire in the early 1800s. Following a series of successful surgical experiments on animals using nitrous oxide, Hickman grasped the implications of his findings but was reluctant to try the gas on his patients. It was a risk Hickman was not willing to take. This decision deprived him of the best advertisement for his discovery, word of mouth from a satisfied customer.

A soft-spoken, unassuming man, Hickman was ill-equipped to promote either himself or his findings. He assumed, incorrectly, that the intrinsic value of his experiments with anaesthesia would speak for itself. A pamphlet self-published by Hickman in 1824, explicitly outlining his success with painless surgery on animals, made its way to the Royal Society, headed ironically by Sir Humphry Davy. You would have thought that the intersection of two similar findings of such magnitude would have produced a 'Bingo!' effect, but Hickman waited for a reply that never arrived. Davy paid scant attention to the insignificant Hickman and, for the second time in his career, Davy consigned humankind to further decades of suffering. The Academy of Medicine in France also ignored the young scientist's pleas for attention.

Hickman's goals were purely altruistic and the almost universal disinterest in his proposals was both hurtful and puzzling. His claim to be able to perform surgical operations 'with perfect safety and exemption from pain' fell on deaf ears. When Hickman met with no success in either London or Paris, he returned home, defeated. He died at thirty, unrecognised. A hundred years later, the Royal Society dedicated a memorial to Hickman, declaring him to be the 'earliest known pioneer of anaesthesia by inhalation', but they were too late to save either the young doctor or the hundreds of people who would have given the earth for the balm of his discovery.

In the early 1830s, the concept of anaesthesia was bubbling, but without a public platform, it stalled. Enter another of our inventors, Samuel Colt, the antithesis of a self-effacing Henry Hickman or a self-important Humphry Davy. In top hat and frock coat, Colt peddled his laughing-gas show at every travelling circus he could find in order to make enough money to finance a Canadian patent for his revolver.

Showmanship came naturally to Colt. On one particular day in Cincinnati, he gathered six American Indians as the main attraction for his show. People shifted uncomfortably, their hands on their guns and canes, as Colt administered laughing gas to the Indians with his home-crafted device. People were ready for anything; anything, that is, except what happened. The Indians dozed off. Desperate to please his audience, Colt turned to a blacksmith who accommodated by sniffing the gas and chasing Colt around the stage. In Colt we have a natural inventor – a risk-taker, an opportunist and a man alive to new ideas – but he showed not a glimmer of recognition in 1832 that he held the key to anaesthesia. As far as Colt was concerned, his eye was on another prize, his revolver.

Another interesting phenomenon was taking place in the 1830s. Young people in America and Europe were enjoying parties at which nitrous oxide, and sometimes ether, was the recreational drug. Like Colt, however, they were looking at the exhilarating effects of the gas rather than the soporific ones. Once again, no one made the intuitive leap.

Next we meet Crawford Williamson Long, a physician in the agricultural community of Jefferson, Georgia. Asked by his students to provide nitrous oxide for their amusement, he offered sulfuric ether instead, which produced the same exhilarating effects.

Recreational 'ether frolics' spread quickly throughout the country. Among the young people who enjoyed the drug were the teenage children of John Collins Warren, the Head of Surgery at Massachusetts General Hospital. The answer to the miseries he inflicted, was, literally, right under his nose.

On March 30, 1842, Crawford Williamson Long administered ether to a patient with a neck tumour. Inexplicably, Long didn't realize the greater implications and only later, after Morton's successful demonstration in Boston in 1846, did Long come forward to claim recognition as the first doctor to administer ether to a patient. Over the four years between 1842 and 1846, Long had operated on three or four patients under ether, but, for whatever reason, it never became public knowledge. Writing later, Long said that he had delayed claiming success until he could be certain that the anaesthetic effect had been produced by the ether and that no other surgeon had used ether prior to Long's own experiments. By 'waiting that eternity for certainty', and delaying publication, Long lost the right to go down in history as the man who introduced painless surgery. In the meantime, patients like a young man in New York, cradled in his father's arms, had a leg amputated without benefit of anaesthesia.

Now we draw closer to the interface of innovation and application. The year is 1844. At a public lecture on the effects of laughing gas, Horace Wells, a distinguished dentist in Hartford, Connecticut, volunteered to go on stage. Noticing that the man next to him had been badly bruised under the influence of the gas but had felt no pain, Wells wasted no time in following the dotted line. Why, he asked, could this not be applied when extracting teeth?

Dr Wells had an inquisitive mind and a strong mechanical bent that resulted in a number of patents for dental instruments. Among Wells' students was William T.G. Morton of Boston. Wells first tried nitrous oxide on himself. Following a painless extraction, he and those around him realised that something extraordinary had occurred. After weeks of trying the new substance on patients, Wells determined to bring his discovery to the attention of the public. William Morton arranged a demonstration at the Massachusetts General Hospital.

At this juncture, it was pretty much down to luck. Wells appeared in the amphitheatre of this famous medical centre, and was introduced as a gentleman who 'pretends' to have something to alleviate the pain of surgery. A demanding and sceptical crowd of students and doctors looked on, among them William Morton. Wells called for a volunteer. A young man came down, the gas was administered and Wells proceeded to extract a tooth. As he raised the forceps, the boy cried out, the students jeered and Wells fled the amphitheatre in disgrace. The patient claimed later to have felt no pain, but it was too late. Wells had been humiliated.

Although Wells later made claim to being the man to first introduce anaesthesia, he has been largely forgotten in popular history. Personal experiments with chloroform later affected Wells' mind and, following an incident in which he threw acid at a streetwalker, Wells was jailed. Wells anaesthetised himself in prison and, just before falling under the influence, slit his femoral artery. He was not to see the letter that arrived a few days later from the Paris Medical Society awarding him the honour as the first to have 'successfully applied the use of vapours or gases whereby surgical operations could be performed without pain'. Horace Wells joins the list of accomplished scientists who had tried and failed to bring not only the fact but the concept of anaesthesia into the foreground.

Now, at last, we encounter William Thomas Green Morton, a dentist practicing in Boston. He had not made any of the great scientific leaps; he had not carried out painstaking experiments. In fact, he was completely different from any of the men we have met thus far.

Forced to leave school at seventeen when his father went bankrupt, Morton left a trail of bad cheques and unpaid debts across the country. The twenty-seven-year-old Morton was wanted by the sheriff in more than one state for fraud and embezzlement. He was reviled by more than one father whose daughter had been left at the altar and the church had seen fit to excommunicate him. Charming and self-confident, he was ready to take a big risk for a big gain. Morton was the consummate shyster with one goal in mind – the pursuit of fortune.

The innovative application of nitrous oxide and ether were known to him through his teacher, Horace Wells, and by way of conversations with other men in Boston. One of those men, Charles Jackson, had suggested the benefits of purified ether over nitrous oxide and later claimed the right of discovery.

When a patient arrived at Morton's office one evening complaining of toothache, Morton decided to try ether. The extraction was a resounding success and Morton, the businessman, was not oblivious to the implications. He had his patient sign a statement, which was witnessed by two assistants, and the next day an account of the painless tooth removal appeared in the Boston papers. Word spread, and over the ensuing weeks Morton administered anaesthesia to a number of patients in his dental office. Morton also took the precaution of maintaining absolute secrecy around the procedure. There was money to be made. On October 1, the day after the tooth extraction, Morton secured a patent attorney, a step unheard of at the time in the medical profession. Prevailing etiquette demanded the voluntary sharing of any significant medical advancement.

To risk a return to the Massachusetts General Hospital, the scene of Wells' humiliation, and try once again to persuade a jaundiced audience that painless surgery was possible, took a certain kind of person. This was a tough group and Morton was far from skilled at the procedure. There were imponderables. Was anaesthetising someone for major surgery the same as putting a patient to sleep for a tooth extraction? Did it require more ether? And when Morton stood ready to demonstrate the wonder of painless surgery, he held in his hand an apparatus he had never used before. There was always the possibility that Morton could inadvertently kill his subject and be prosecuted for manslaughter, so little did he or anyone know about the process.

When a letter arrived from the hospital inviting Morton to demonstrate anaesthesia, he had only one day to prepare, but his was not a nature to turn away from risk associated with opportunity. In terms of actual innovation, just two things made the process his own: the addition of orange to the ether to improve the scent and a new apparatus for administration, one that diverted exhaled breath. This apparatus was still under construction on the morning of the surgery. There would be no time to test the new piece of equipment before the demonstration.

When Thompson arrived to pick up the new applicator, it wasn't ready. The minutes crept by as the audience in the amphitheatre waited impatiently. At last Morton strode onto the stage and approached the young patient, one Mr Gilbert Abbott, who had been strapped into the surgical chair. Tiers of seats rose above the stage; sun streamed in through the glass dome. Standing room only. Hushed silence.

'Are you afraid?' Morton asked.

'No,' answered the patient. 'I will do exactly as you tell me.'

In this very gesture, the instilling of confidence, Morton may have turned the tide. His patient, relaxed and willing, inhaled the ether and remained insensible through the removal of his tumour. The onlookers cheered. Ether had its debut at the hands of a lucky innovator, a man who could barely put together a coherent sentence. What Morton had done was to seize the moment.

The next day ether was successfully applied again. Morton wasted no time in applying for a patent and mounting an advertising campaign. A group of eminent men was asked to come up with a name for the process. Oliver Wendell Holmes suggested Anaesthesia.

In November 1846, Annie Moran, a twenty-year-old girl, was anticipating the grim prospect of an amputation. She was the battleground on which the validity of Morton's patent would be fought. The stakes were high. The Massachusetts General Hospital, refusing to be blackmailed, decided to operate without anaesthesia unless Morton revealed his formula.

The patient was rolled into the theatre and tied down. Morton, standing outside the amphitheatre doors, finally agreed to reveal the process and Annie was spared. When Morton realised that he could not prevent infringement of his patent, he was wise enough to try a different approach. A petition to the United States Government for $100,000 in payment for his innovation languished in committee for twenty years before being dismissed. His claim to have introduced anaesthesia was challenged on all sides.

Gradually, overrun with lawsuits and failing to realise the financial rewards he had hoped for, Morton's fortunes dwindled. Riding through New York's Central Park with his wife on a hot summer's day, he leapt from his carriage and ran to the lake, where he submerged his head in the cool water. Returning to the buggy, he drove a short distance before jumping out once again, this time to hurdle a fence. William Morton died a short time later, probably of a cerebral haemorrhage.

A pass at the moon

JONAS SALK

Born: October 28, 1914, New York City
Died: June 23, 1995, La Jolla, California

*Intuition can be improved by reason, but reason alone,
without intuition, can easily lead the wrong way.
Both are necessary. I might have an intuition
about something, so I send it over to the reason
department. Then, after I've checked it out in the
reason department, I send it back to the intuition
department to make sure that it's still all right.
That's how I work.*

Jonas Salk

Do only that which makes your heart leap.

Dr Gregg to Jonas Salk

OFTEN, IT STARTED WITH A SLIGHT HEADACHE, perhaps some nausea, a low fever. Within twenty-four hours, paralysis might set in and a healthy, active child was suddenly facing a life on crutches or with leg braces or, at worst, in an iron lung. The lucky ones recovered. Some, with light symptoms, never even knew that the dark shadow of polio had brushed their lives.

Instead of looking forward to long, happy summer days, parents waited anxiously as the heat settled in and, with it, the terror of infantile paralysis. There was no predicting where or when it would strike next or who would be felled. The rich, including President Franklin Roosevelt in his country home in Canada, were as vulnerable as the family living in a tenement in Brooklyn. Between 1950 and 1955, some twenty-five thousand cases of polio occurred annually, with fifty thousand cases recorded in 1953. A nurse recalls a young boy weeping because he was completely paralysed, unable to lift a hand even to kill himself. 'Can't you do something?' was all she could say as the doctors walked past.

The first major polio epidemic, claiming more than seven thousand lives, occurred in 1916, two years after Jonas Salk was born to uneducated Russian Jewish parents in New York. His mother, overprotective and domineering as she pushed Jonas to excel, instilled in her son, the doctor, a lifelong resistance to being controlled. When Salk decided to become a lawyer, his mother advised against it because, as Salk later explained, if he could never win an argument against her, what kind of a lawyer would he make?

From the earliest days at medical school and later, during his residency, Salk felt sure that scientific research was where he belonged. His rejection by a number of prominent research facilities and medical schools, in part a reflection of the strong anti-Semitism that prevailed at the time, was a disappointment, but led him instead to a position working on influenza vaccine. Much of the groundwork was laid there for his later work with polio.

In 1948, Jonas Salk was offered the opportunity to start his own lab at the University of Pittsburgh Medical School, and he jumped at the chance. When asked why he would align himself with an institution of such mediocre reputation, Salk answered that he had 'fallen in love'. He explained later, 'What I was in love with was the prospect of independence.'

Perhaps nothing was as important to Jonas Salk as the freedom to pursue his own way of thinking and working. Rules and regulations did not sit well with him; he preferred action to filling out forms or sitting through long committee meetings. Jonas Salk was also a perfectionist. Donna Salk confessed in an interview that when her husband cleaned the stove, a process that took all day, he pulled it out into the middle of the room, disconnected the gas cord and cleaned the heads of the screws with a toothpick.

A year after Salk arrived in Pittsburgh, the director of research for the National Foundation for Infantile Paralysis approached him about participating in their poliovirus-typing programme, and Salk jumped at the chance to fund his laboratory. The Foundation, conceived by polio victim President Franklin Roosevelt and richly funded through the country-wide March of Dimes movement, was a powerful force in the search for a cure for polio. The head of the Foundation, former Wall Street lawyer Basil O'Connor, was willing, like Salk, to cut through the scientific hierarchy when necessary.

Most scientists were loathe to become involved in the process of virus typing, a laborious but necessary step in the development of a vaccine. They were happy to leave it to a junior researcher. But where others saw drudgery, Salk saw opportunity. He was happy to be in control of his own laboratory and welcomed the hundreds of thousands of dollars that poured in from the Foundation. He also took advantage of the time to set up his lab, build a team and become familiar with the poliovirus.

In 1948, John Enders of Harvard and his colleagues Thomas Weller and Frederick Robbins announced the revolutionary tissue-culture technique that was to clear the way for vaccine production. Now the polio virus could be cultivated in unlimited quantities in a medium other than foreign nervous tissue, such as the monkey's spinal cord. Enders refused to launch his own vaccine project based on his discovery, claiming that his lab wasn't set up for such an undertaking, but Jonas Salk didn't delay. By the time other researchers had invested in the critical new tissue techniques, Jonas Salk, who had moved immediately to adopt the new concept and move it forward, was way ahead of them. By March 1953, Salk's Virus Research Lab was producing poliovirus in cultures.

The key principle at the heart of the vaccine debate was the notion of using a killed virus vaccine versus a live vaccine – live virus that had been 'attenuated' or weakened sufficiently to inject safely and promote the production of antibodies. Among those favouring a live vaccine was Albert Sabin. Salk held that a killed vaccine, one that had been chemically neutralised, could spur an immune system to produce antibodies without the danger of an actual infection.

The National Foundation's Committee on Immunization met in New York in January 1953. At the morning session, the researchers agreed that the chance of producing a polio vaccine any time soon was negligible. Remarkably, there was no sense of urgency in spite of the devastating epidemic of the summer of 1952, when news of the spread of polio had been the subject of nearly every front-page headline for months on end.

The tone changed after lunch, when Jonas Salk, a junior member of this elite gathering, rose to speak in a calm, methodical manner about his preliminary vaccine tests with children at the Polk School for the Retarded and the Watson Home for Crippled Children. News of these trials came as a surprise to almost everyone at the meeting, a miracle considering the number of people involved in the process. None of the subjects, Salk reported, had suffered negative consequences from immunisation; all had developed antibodies. The next step would be testing on a much wider scale and Salk wanted the Committee's support.

Dr Albert Sabin, at work on his own live vaccine, was vehement and condescending. This was absurd, he countered; it would take ten to fifteen years of additional research to develop a completely safe vaccine. At the back of Sabin's mind, as perhaps with other scientists, was the memory of two unfortunate colleagues, John Kolmer and Maurice Brodie, whose careers had been ruined when the vaccine they introduced in the 1930s proved to be faulty. Jonas Salk wasn't one to shy away. 'Risks always pay off,' he was to write later. 'It is there that you learn what to do or what not to do.'

'Not infrequently,' Salk explained, 'people in the lab would tell me that something hadn't worked. And I would say "Great, we've made a great discovery."'

For Basil O'Connor, head of the Foundation, the question was fundamental. Was it better to go ahead with something that hadn't been perfected and save thousands of children or wait more than a decade for something certain? To await certainty, he believed, was to await eternity. Calculating that those twenty years would translate into a million cases of polio, of which twenty thousand victims would be paralysed, O'Connor decided to take the chance. Sabin predicted a hideous fiasco leading to professional suicide for Jonas Salk.

Salk was very much aware of the antipathy emanating from the scientific hierarchy, but he chose to move ahead on his own. Thin and studious, peering out from behind thick glasses, he had never been 'one of us'. Now his status as an outsider worked in his favour. He wasn't paralysed, as many of his fellow scientists were, by the need to preserve the status quo. On the other side of the table sat people who were trying, at any cost, to hold the ground that provided security and defined their reputation. Many of them had invested years of work on their own vaccines. Salk's crime was to question the way people looked at things – that was his nature. He had also bypassed the age-old scientific order: research and publish before moving into action. As one friend put it, 'In the time it took others to proceed step by step, Jonas had made several passes at the moon.'

There was nothing more unattractive than a scientist who went after the spotlight. 'Dear Jonas,' wrote Albert Sabin 'This is for your information, so that you'll know what I am saying behind your back. And incidentally, this is also the opinion of many others whose judgment you respect. Love and kisses, Albert.'

In March 1953, Salk injected about six hundred additional subjects, including his three sons. In that same month, Basil O'Connor, seeing that the Committee on Immunization would refuse to move forward, formed a new Vaccine Advisory Committee whose members had no stake of their own in the race for a vaccine. To bypass the leading authorities in viral research, all of whom cautioned against a field trial, was a courageous step.

To prepare for the 1954 nationwide field trials, forty-year-old Salk was, meanwhile, working out details and preparing for mass production of his vaccine. He saw the vaccine as a moving target, which could and should be constantly improved. Production specs were tweaked from day to day. The drug companies, frustrated by the changes, were tempted to cut corners, but a comprehensive testing process screened out vaccine that wasn't acceptable. All this took place amidst a barrage of media coverage, most of it positive, some of it venomous. Some colleagues accused Salk of self-aggrandisement, others, impatient for a product, accused Salk of being timid and unwilling to relinquish control. The attacks were fierce.

'Good evening, Mr and Mrs America, and all the ships at sea,' piped columnist Walter Winchell over the radio networks. 'In a few moments,' he continued, 'I will report on a new polio vaccine and it may be a killer.' Winchell went on to describe the little white coffins that had been stockpiled around the country in preparation for the children who would die during the field trials. It is remarkable that Jonas Salk moved forward with seeming assurance and determination in the face of such opposition. Confronting this onslaught, the members of the Vaccine Advisory Committee were terrified, but they put their reputations on the line and voted unanimously to go ahead with nationwide field trials.

If scientists were against this unprecedented trial, parents couldn't wait to line up their children. Their fear of the disease far outweighed concern about possible side effects. One woman remembers her mother signing the consent form then collapsing over the table in tears of relief. In April 1954, just in time to beat the epidemic season, the first six hundred and fifty thousand children received their injections. The almost two million schoolchildren in forty-four states who participated in the trials became known as Polio Pioneers. More than one million two hundred thousand needles were required. Jonas Salk stood to be a world hero or suffer as one of the most public failures in history.

On April 12, 1955, Dr Thomas Francis prepared to reveal the results of the field trial, meticulously derived from carefully assembled data. The gathering in Ann Arbor, Michigan was huge. People sat huddled around their radios; department stores set up loudspeakers. Then came the news: 'The vaccine works. It is safe, effective and potent.'

If Jonas Salk had been an outsider among his colleagues before his success, he was even more so now. His public accolades failed to acknowledge the other scientists whose work had played a part in isolating the vaccine. Salk's long-suffering laboratory staff, all of whom had contributed hugely to the project, were barely remembered in Salk's remarks. Enders' associate, Thomas Weller, declared the Salk vaccine 'one of the biggest scientific farces ever foisted on the public'.

'The worst tragedy that could have befallen me,' Salk said later, 'was my success. I knew right away that I was through, cast out.'

Fifty years later to the day of Dr Francis' announcement, Jonas' son, Dr Peter Salk, stepped up to the microphone at the event commemorating that day in Ann Arbor and dealt with unfinished business. 'Nothing can be done to turn the clock back fifty years,' he said, 'but everyone please accept my expression of recognition. The vaccine was not the accomplishment of one man, it was the accomplishment of a dedicated and skilled team.' Peter Salk then read the list of team members. No one really knows why Jonas Salk failed to thank his team that day, but many of his associates left the ceremony in 1955 in tears.

April 22, 1955 dawned a beautiful day in California. Josephine Gottsdanker took her daughter Anne for a ride and they stopped for ice cream. Then, during the afternoon, Anne developed a headache and experienced nausea. By the time they arrived at the hospital, Anne couldn't move her legs. Two days earlier, Anne had received the Salk vaccine; now she had polio. Her brother, who had also been vaccinated, remained healthy.

Anne was one of nearly forty thousand children in the course of that tragic spring who contracted polio in spite of vaccinations. Two hundred were paralysed; eleven died. In a double-blind field trial, some had received placebos and the vaccine was admittedly only 60% effective, so some cases were to be expected, but a pall was cast over the celebrations that marked the end of polio's reign of terror. Salk described it as a tragic disappointment. Investigation revealed that vaccine produced by Cutter Laboratories, released without benefit of the stringent testing imposed during the field trials, had been faulty. And in the way that people have of moving quickly onto the next thing, the public displayed a short memory for the terrible anguish of the worst polio epidemics. The numbers opting for vaccination dropped to 35%. When Sabin's oral vaccine was introduced in 1961, after field trials carried out in Russia, it quickly overtook Salk's product, although Jonas Salk forever stood as the man who had conquered polio. There were isolated cases of polio directly related to Sabin's vaccine as well, and the debate went on.

Professional recognition was never to come to Jonas Salk. There was no Nobel Prize, no membership in the elite Academy of Sciences. When asked why, other scientists explained that Salk was a synthesiser, that he had contributed nothing new to the field. All he had done was to pull together existing technologies and run with it. That thinking extended to the legal opinion that the Salk vaccine was not patentable. When asked by Edward R. Murrow whether he had applied for a patent, Salk replied that to patent the vaccine would be like trying to patent the sun.

After the introduction of the polio vaccine, Jonas Salk turned his attention to the establishment of an independent intellectual centre in La Jolla, California. The citizens of La Jolla were not pleased. One of Salk's speeches was halted by demonstrators holding a sign that read, 'Don't kill dogs in the night.' Salk left the hall without delivering a word. Once the magnificent Salk Institute overlooking the Pacific had been finished, Salk found that the great minds he attracted came attached to equally big egos. Rather than appreciate their good fortune, many of these scientists looked upon their benefactor as nothing more than a technician. There was no utopia here.

When Salk's twenty-eight-year marraige fell apart, he wed Françoise Gilot, an artist and long-time mistress of Pablo Picasso. Many people would have broken down under the difficult circumstances, but the philosopher residing in Jonas Salk prevailed, lending perspective to events that would otherwise have been experienced as tragedies. 'Number one in life,' he wrote, 'is to have a purpose. Your purpose can be different at different times in your life, but take good care of that purpose and let it be your guide.' Contrary to what many claimed, Jonas Salk had never sought fame. He had simply wanted to make a difference and in that he succeeded absolutely.

In La Jolla, Salk carried out studies on the immune system and its role in diseases such as cancer and multiple sclerosis. He was involved in testing a noninfectious polio vaccine providing life-long immunity in a single dose and, during the 1980s, studied the AIDS virus in partnership with his son, Peter. Salk also wrote extensively, in particular on human evolution and a new form of human consciousness. He was involved in a wide variety of philanthropic organisations and contributed generously both time and money.

A column in the Sunday *New York Times* remarked on the death on May 28, 2008 of a sixty-one-year-old woman who had spent all but three years of her life in an iron lung. For more than six decades, while her peers grew up, married and had children, danced and climbed mountains, she had lain on her back, facing the ceiling, observing the world around her with the help of a mirror. Her world was an iron tube on which she depended for her very breath. But for Jonas Salk and his willingness to take a chance, there would have been many hundreds more like her.

Dr Jonas Salk died of heart trouble at the age of 80 at Green Hospital in La Jolla.

The way to an artificial heart

PAUL WINCHELL

Born: December 21, 1922, New York City
Died: June 24, 2005, Moorpark, California

*"Ventriloquism is closely related to magic.
It's all about misdirection."*
Paul Winchell

"Hallo, Piglet. This is Tigger."
*"Oh, is it?" said Piglet, and he edged round to the other side
of the table. "I thought Tiggers were smaller than that."*
"Not the big ones," said Tigger."
A. A. Milne, *The House at Pooh Corner*

WHEN YOU HEAR THESE WORDS AND OTHER INSPIRATIONAL comments
from Tigger to his Winnie the Pooh friends in the Disney cartoon adaptations, you
may well be listening to the voice of Paul Winchell, who immortalised this bouncy
tiger in his voice-overs. He also gave life to two famous dummies, the dim-witted
Knucklehead Smiff and his wooden sidekick, the sassy Jerry Mahoney. Winchell and
his friends were a popular item on American television in the 1950s and '60s.

Paul Winchell, born Pinkus Wilchinski, later shortened to Wilchin, grew up in
a cold-water flat on the Lower East Side of New York. He contracted polio at the age

of six and fought hard to bring new vigor to his atrophied legs. He later wrote that his mother was intolerant of his disease and beat him frequently. While lying in bed, Winchell occupied himself by sending off for coupons and give-aways from magazines, one of which was 'Be the Life of the Party: Throw your Voice.' He also relieved his loneliness by listening regularly to the Edgar Bergen show, featuring Bergen's mannequin Charlie McCarthy, on 1920s radio in America. Overcoming his family's opposition, his shyness and a childhood stutter to become a ventriloquist, Winchell introduced his alter-ego, the puppet Jerry Mahoney, on radio in 1936, debuting on NBC television in 1947. Over the following years, Winchell was watched by millions of dedicated fans every week on national television.

With an impressive list of voice-overs and guest appearances, Paul Winchell had a career that would have fulfilled anyone's dreams, but ever the inventor of personalities, Winchell also applied for thirty patents on items as diverse as battery-operated gloves, an invisible garter belt, a flameless cigarette lighter, a retractable fountain pen and a freezer-interrupt indicator that enabled people to determine whether their food had defrosted when the electricity went down. Some claim that Paul Winchell was the first to conceive the idea of the disposable razor. When skeptics impressed him with the idea that, 'no one would buy a product and then throw it away,' Winchell dropped the idea, only to see others make billions.

In the mid-1950s, Winchell took pre-med courses at Columbia University, stating that 'It wasn't until I was 35 that it dawned on me that I'd missed my education.' Winchell took courses in acupuncture and hypnosis, attributing his place as one of the top students in the group to his photographic memory. Soon, Paul Winchell was in demand for hypnotherapy work in post-operative cases. When his own son had his tonsils removed, Winchell employed hypnotic suggestion and was allowed into the operating room to observe the surgery.

When Winchell arrived home from his son's surgery, there was a call waiting from his agent. Would he, the agent asked, be willing to participate on the *Arthur Murray Dance Party* show on prime-time national television? First prize was a new Buick.

Paul Winchell, dancing the jitterbug with a Murray dance instructor, defeated Ricardo Montalban (who performed a tango) and won the car. The Murrays threw a

cast party in Westchester, and Paul Winchell drove up in his new Buick. At the gathering he met Hank Heimlich, the father of the Heimlich manoeuvre. At the time, Heimlich was chief resident at the Montefiore Hospital in the Bronx section of New York and he invited Winchell to scrub in on some operations. After a day observing in the operating room, including open-heart surgery on a young girl, Winchell was profoundly impressed by the daring and innovativeness of the procedures. He was to return often as an observer.

When, one day, a patient died on the operating table, Winchell went to his friend Doctor Heimlich and asked whether, perhaps, an artificial heart, with its own power source, might keep a patient alive during a critical period. 'He looked at me,' wrote Winchell, 'and smiled. "You build your own dummies," Heimlich said to me, "So why don't you make a model of your idea?"'

That was all Winchell needed to hear, and off he went to work on his model heart.

'Odd as it may seem,' he wrote, 'building the heart wasn't that different from building a dummy; the valves and chambers were not unlike the moving eyes and closing mouth of a puppet.' True to his word, Heimlich was always ready with advice and suggestions for corrections and improvements. 'At last, after examining all the changes,' Winchell recorded in his autobiography, 'Hank looked at me and said, very simply, "If this were my idea, I would get it patented."'

Winchell duly applied for a patent for his artificial heart in 1956; the patent was finally granted in 1964. A workload that included time spent on his invention and a Saturday morning TV show precluded his continuing at medical school, so Winchell began instead to concentrate on his hypnotherapy.

Not long after the examiner had granted Winchell a patent for his artificial heart, a letter arrived from a Doctor Kolff at the University of Utah saying that they also had attempted to patent an artificial heart, but had been refused on the basis of Winchell's patent as 'prior art.' Kolff requested a meeting. Winchell arrived at the University of Utah to be shown a cow called Lord Tennyson, that had been keps alive for months through the use of an artificial heart.

Doctor Kolff asked that Paul Winchell donate his patent to the university. Although he asked for time to consider the request, Winchell confessed that 'it was never my intention to make money on the device because I always considered it as a contribution to humanity.' The patent was duly turned over to the university. Very quickly, the attitude towards Winchell changed and communication from Dr Kolff and the university ceased. Once his patent was out of the way, all mention of Paul Winchell in connection with the artificial heart ceased. When anyone called the university to enquire, they were told that Winchell had nothing to do with the artificial heart project.

Winchell was also deleted from history when Metromedia erased 288 tapes of Winchell's children's television series *Winchell Mahoney Time*, produced live on air between 1964 and 1968. It was Metromedia's answer to a dispute with Winchell over syndication rights. In 1986, Winchell won a lawsuit against Metromedia to the tune of $17.5 million but the heart of his best work on television was gone forever.

When Paul Winchell died, one of his friends wrote about the private man, 'the man who was not a performer,' a man plagued by 'inner demons,' describing him as a man who had a severe stutter, suffered from polio and the rejections of a cold and cruel mother. Paul Winchell was a troubled, complicated man, full of contradictions. When he died, his daughter April, one of five children and step-children wrote:

'I got a phone call a few minutes ago, telling me that my father passed away yesterday. A source close to my dad, at least closer than I was, decided to tell me himself, rather than letting me find out on the news. My father was a very troubled and unhappy man. If there is another place after this one, it is my hope that he now has the peace that eluded him on this earth.'

Playing with art

ALBERT COOMBS BARNES

Born: January 2, 1872, Philadelphia, Pennsylvania
Died: July 24, 1951, Philadelphia, Pennsylvania

He was jealous, though he did not show it,
For jealousy dislikes the world to know it.

Lord Byron

ALBERT COOMBS BARNES, THE SON OF A PHILADELPHIA LETTER carrier, trained as a medical doctor, but had neither the temperament nor the time to see patients. What did interest Barnes was the prospect of creating a product that would be easy to produce and market and which could potentially make him a rich man. Invention, for Barnes, was nothing more than the key to financial freedom.

The magic potion, developed in 1902 in partnership with a gifted research chemist, Hermann Hille, was Argyrol, a compound of silver and protein used as a topical ophthalmic antiseptic in the pre-antibiotic era. Argyrol dominated the market for almost the whole first half of the twentieth century. Working on the same theory that John Stith Pemberton (see page 29) had previously adopted in the nineteenth century to protect his formula for Coca-Cola, Barnes trademarked but never patented Argyrol. The magic mix was kept under wraps.

Bypassing more traditional marketing routes, Barnes sold Argyrol directly to physicians, both in America and overseas, making good use of testimonials from eminent scientists. By 1904, sales had soared to $100,000 on an initial capital investment of $1600. Argyrol, easy and inexpensive to manufacture, was more or less a two-man operation. It was also a gold mine. Five years after the introduction of

Argyrol to the market, Albert Barnes, aged thirty-five, was a millionaire.

As he was to do with nearly everyone over the years, Barnes soon fell out with Hille, whom he bought out for the sum of $350,000. From that point on, Hille, who went on to be successful in his own right, was erased from Argyrol's history. Leaving his factory in the charge of trusted employees, Barnes moved on to other things.

Plagued by boredom and full of energy, Albert Barnes had achieved his first goal in life, to be very rich by the time he was forty. Settling with his wife, Laura, on Philadelphia's Main Line, home to that city's most distinguished citizens, he built a large residence and established a household worthy of his new position, complete with maids, butlers and cooks and a cellar stocked with fine wines. When none of the trappings of wealth won him a place among the people who prized breeding over ingenuity, Barnes was offended and an adversarial relationship took root.

Rejected by his new neighbours, Barnes turned his sights to the world of art, where his vision manifested itself in a different way. Learning quickly to assess what had value and what did not in the art world, he sent a representative to Paris to buy what would become the nucleus of the Barnes Collection: Cézanne landscapes, a Degas, a small Renoir portrait of a girl reading (for which he paid $1,400) and a Van Gogh. In December 1912, Albert Barnes travelled to Paris, where he purchased his first two works by Henri Matisse. By 1923, Barnes owned fifty Cézannes. It took a while for Barnes to adjust to the 'new' art, but he soon understood and appreciated the modern painters. From that point onwards, Barnes made his own purchases.

As his collection grew, Barnes established a Foundation at his home outside Philadelphia, dedicated to the instruction of art appreciation and philosophy. The building, designed to house his art collection and serve as a centre of learning, was constructed from nine hundred tons of limestone imported from Europe. Barnes had a soft spot for serious young students, particularly when he sensed the opportunity to mould minds and sensibilities, and he could be very generous; many a student went to Europe to study courtesy of Dr Barnes. 'It is that plain, ordinary person,' Barnes wrote, 'with little schooling, whom we want to teach to use the qualities of mind, heart and soul, with which he has been endowed by nature, in such a way that he will be able to understand what he thinks and the artists have done. That is the main idea of the Foundation.' Unfortunately, the conservative world of art on the East Coast did not share Barnes' enthusiasm for the new artists. Many of the Impressionists and Post-Impressionists were denounced as 'insane' and 'degenerate'. One commentator remarked that the artists, many of whom were foreigners who had moved to France, 'should have stayed home'.

Much of Barnes' antipathy for his neighbours stemmed from the public's reaction to the first showing of his collection at the Philadelphia Academy of Fine Arts in 1923. Cézanne and Matisse were considered wildly modern and the Philadelphia papers accused Barnes of subverting public morals. Circulating through the crowd, Barnes heard himself referred to as a freak and worse. Barnes' angry isolation from those around him became increasingly acute. His pride bruised, angry and disappointed, he never forgave the powers that be, and for the rest of his life most members of the establishment were banned from his collection.

In one particularly acrimonious exchange, Barnes approached his alma mater, the University of Pennsylvania, outlining a programme of cooperation that would have resulted in the Barnes collection becoming university property. When the provost failed to respond to Barnes' letter, the opportunity to own an extraordinary group of paintings that included — among many others — sixty Picassos and more than a hundred Renoirs was lost to the university forever.

When dealing with Albert Barnes, it was a definite advantage to be disadvantaged. The founder of New York's Museum of Modern Art arrived at the Foundation gates at 8:00 a.m. as instructed, but no one was there to let her in. When T.S. Eliot requested a chance to view the collection, Barnes' answer was a very clear: 'Nuts'. The famous architect Le Corbusier, unable to make the assigned appointment time, wrote a conciliatory letter which was returned inscribed with the word 'Merde'.

'I was familiar with your reputation in Paris as a boob to whom the dealers could sell any worthless picture,' Barnes wrote to one distinguished Philadelphia collector. A college professor requesting a tour was told that she would have to take an intelligence test first. The privileged were the enemy. Walter P. Chrysler, Jr's request for a visit was answered by a fictitious secretary, one 'Peter Kelly', who advised that such a visit would be 'impossible', as Dr Barnes had given 'strict orders not to be disturbed during his present efforts to break the world record for goldfish swallowing'. Some letters, including one to Alexander Woolcott, a famous journalist and drama critic, were signed with his dog's name, Fidèle de Port Manec'h. Other replies were signed 'The Janitor'.

Barnes identified with the underdog as avidly as he detested the establishment, and there were exceptions to his sociopathy. Barnes was very fond of his pals at the local fire department, of which he was an honorary colonel. Fire Chief Albert Nulty, a Foundation trustee, was among the handful of loyal and trustworthy employees and students. These people saw another side of Albert Barnes, sometimes playful, often generous and punctuated with great acts of kindness. More common, however, was his reputation for irascibility. The *Saturday Evening Post* characterised him as 'a combination of Peck's Bad Boy and Donald Duck', a man who has 'made capital out of rudeness'. A group of local ladies, invited for a formal tea, were treated to a unique view of the doctor as he walked naked through their midst on the way back from the pool. His wife, it seems, had asked him to wear a robe on his way out for a swim, but had not covered her bases for the return trip.

The Barnes' country home, Ker-Feal, 'Fidèle's home', named in honour of their beloved dog, provided an important retreat for Barnes and his wife. The black and white mongrel, brought to America from Brittany, shared his master's bedroom. Above Fidèle's cushion hung an eighteenth-century map of his native Brittany, 'in case he felt lost'. Above the doctor's bed hung a drawing of Fidèle's village, Port-Manec'h.

In July 1951, at the age of eighty, Barnes was driving his Packard back to Philadelphia, his stalwart companion Fidèle by his side, when he ignored a stop sign and was hit broadside by a ten-ton trailer truck. The only people who commemorated his death were his friends at the fire department, who lowered their flag to half mast and draped their headquarters in black bunting.

Control of the Barnes collection, one of the world's finest, comprising Impressionist and post-Impressionist art, African works and craft from the American West, was accorded to Lincoln University, a small historically black college located in Chester County, Pennsylvania. In the 1950s, Albert Barnes amended the by-laws such that upon his death, Lincoln's board of trustees would nominate four of the five trustees of his foundation. On June 5, 1951, Lincoln University awarded Barnes the honorary degree of Doctor of Science. The collection includes, among many other works, 175 Renoirs, 65 Matisse canvasses and 66 Cezannes.

As for Argyrol, in July 1929, three months before the stock market crash, Albert

Barnes sold his manufacturing company and the Argyrol formula to the Zonite Products Corporation for $6 million. The money went straight into the bank. In 1961, the art collection was valued between $75 and 100 million.

I speak but no one will listen

IGNAZ PHILIPP SEMMELWEIS

Born: July 1, 1818, Buda, Hungary
Died: August 13, 1865 Vienna, Austria

All progress depends on the unreasonable man because reasonable men accept the world as it is, while unreasonable men persist in adapting the world to them.

George Bernard Shaw

AS AN ASSISTANT AT VIENNA GENERAL HOSPITAL'S OBSTETRIC CLINIC in the 1840s, Dr Ignaz Semmelweis was struck by the fact that in this medical teaching facility the maternal and infant mortality rate was more than thirteen per cent. Another clinic, dedicated to the instruction of midwives, had a much lower rate of two per cent. Puerperal fever was the killer and the assumption was that this disease was an inevitable part of hospital-based obstetrics. Ironically, at a time when most women delivered at home, where mortality rates were negligible in comparison, the majority of maternity-ward patients were the poor and the indigent.

What, Semmelweis asked himself, was making the difference?

Semmelweis was on holiday in Venice when his friend and pathology professor, Jakob Kolletschka, cut his finger during an autopsy and died. The pathology report on Kolletschka revealed a condition not unlike that seen in the women dying from puerperal fever and Semmelweis, thinking outside the box as innovators and inventors do, drew a dotted line between the contamination present in cadavers and the occurrence of puerperal fever in patients.

Inferring that medical students were carrying infection on their hands between the autopsy room and their obstetric patients – something that would not be true of midwifery students – Semmelweis instituted new procedures in 1847. Students and doctors were, from that time onwards, required to wash their hands in a solution of carbonated lime when moving from work in the pathology lab to the labour ward. A year later, he extended the prophylaxis to instruments and went on to mandate that bed linens be changed regularly.

Ignaz Semmelweis was not the first to reach these conclusions. The eighteenth-century Scottish physician Alexander Gordon asserted that obstetricians should wash their hands and clothes before treating patients and American physician and essayist Oliver Wendell Holmes, whose life spanned most of the nineteenth century, wrote in 1843 about his belief that puerperal fever was spread when doctors did not wash properly. Holmes died of septicaemia shortly after writing his essay. Semmelweis took things one step further in instituting procedures and obtaining conclusive evidence in a controlled environment. 'God only knows,' he wrote, 'the number of women whom I have consigned prematurely to the grave.'

Even when the mortality rate in the teaching clinic dropped from twelve to two per cent in a month, there was little support for Semmelweis' new theories of hygiene. The head of the clinic, Johann Klein, who maintained that puerperal fever was non-preventable, resented the challenge to his leadership and declined to reappoint Semmelweis to his post.

Funded by the Vienna Academy of Sciences, Semmelweis moved on to animal experiments. By 1850, he was at last prepared to present his findings to the medical community and delivered a lecture to the Association of Physicians in Vienna. Opposition continued to be strong. When, a few months later, Semmelweis was appointed as an instructor in midwifery, he was restricted to mannequins rather than patients. Humiliated, he abandoned Vienna's medical establishment in favour of an appointment as head of the maternity ward of Pest's St Rochus Hospital. When implementation of disinfecting procedures reduced birth clinic mortality rates to less than one per cent, the Hungarian government ordered that Semmelweis' prophylactic methods be introduced in all hospitals. As news of his work spread, he was offered the chair of obstetrics in Zurich; the authorities in Vienna remained hostile. 'It is time,' wrote the editor of the medical journal there, 'to stop this nonsense of hand washing with chloride.'

In 1861, Semmelweis finally published his findings, but his book was poorly written and received a number of negative reviews. At a conference of German physicians and scientists, most participants spoke out against the Semmelweis doctrine. In anger and frustration, he lashed out in print, further alienating himself from the medical community.

By 1863, the conflict had taken its toll on his spirit and Semmelweis was, alternately, apathetic and enraged, thwarted, as he saw it, in his appointed role as the saviour of new mothers. At the age of forty-seven, Semmelweis was incarcerated in an asylum, where he died two weeks later. Ignaz Semmelweis was apparently the victim of a sepsis ironically similar to puerperal fever, resulting from an infected finger. Evidence later came to light that Semmelweis had died from injuries received when he was beaten by asylum personnel.

'When I look back upon the past,' wrote Semmelweis in the foreword to his book, 'I can only dispel the sadness which falls upon me by gazing into that happy future when the infection will be banished. The conviction that such a time must inevitably, sooner or later, arrive will cheer my dying hour.' The Medical University of Budapest, where he worked, is now named after him.

The Wizard

THOMAS ALVA EDISON

Born: February 11, 1847, Milan, Ohio
Died October 18, 1931, West Orange, New Jersey

School! I've never been to school a day in my life!

Everyone steals in industry and commerce.
I've stolen a lot myself.
The thing is to know how to steal.

Thomas Edison

HERE IS THOMAS EDISON, with a cigar in his hand and a twinkle in his eye, describing how he rose from nothing to become a personality so widely known that an envelope bearing only his picture reached him in Orange, New Jersey. In many ways he is king among inventors, claiming not one or two, but dozens of seminal inventions. And along with that, he was ruthless, using everything from the patent process to other people to his advantage.

Edison was also a broker. He was a showman, ready to bend the truth a little when necessary. Here was a man willing to electrocute an elephant to prove his point! Admire him? Yes. Like him? Perhaps not, but I don't think he would have cared. Part of his strength was in being eccentric, in living life in his own way, and that freedom served him well.

It is true that Edison, an avid reader, had very little formal education, but his mother, a former schoolteacher, spent a lot of time encouraging Thomas, her seventh child. Edison's father, who dreamed big dreams but accomplished little of substance, endowed his son with one enormously important quality, eternal optimism. Edison was never deterred by failure, but accepted it as part of the trial-and-error inventive process. 'I have not failed,' he was fond of saying, 'I have just found ten thousand ways that don't work.' He chided, 'It is absurd to say that because I can see no possible solution to the problem today, I may not see one tomorrow.'

Thomas Edison was twelve when he first climbed aboard the wooden cars of the Grand Trunk Railroad with a stack of newspapers and a box of candy to sell to the passengers. His days began at 7:00 a.m. and finished at 8:00 p.m. when the train pulled back into his hometown of Port Huron, Michigan. Spare time aboard was spent experimenting with the chemicals he collected and stored in the baggage car. When some of the bottles broke and started a fire, the conductor threw out Edison's precious collection. Not to be deterred, Edison turned his hand instead to publishing a newspaper, which he sold to passengers on the train.

Fascinated by the telegraph, Thomas Edison taught himself the Morse Code and found a job in the Port Huron telegraph office, where his tasks included topping up of the acid in the primitive batteries. Never one to leave well enough alone, Edison experimented with the mixture, blowing the office and everything in it to smithereens in the process. Observers reported seeing the little wooden building explode and three small boys flying out of the door.

Bored by routine but willing to work hard, intensely curious, endowed with a vivid imagination and the absolute confidence that he could solve any problem, young Thomas Edison travelled from town to town, picking up mechanical expertise and a good knowledge of the telegraph.

Edison is quoted in 1917 as saying that for most of his life he had refused to work at any problem unless the solution could be put to commercial use. It was certainly his ambition to make money through his inventions, but the greatest goal for Edison was a model workshop, not a rich lifestyle.

'My greatest luxury would be a laboratory more perfect than any we have in this country,' Edison told an interviewer in the 1860s. 'I want a splendid collection of material – every chemical, every metal, every substance, in fact, that may be of use to me, and I hardly know what may not be of use. I want all this at hand, within a few feet of my house. Give me these advantages and I shall gladly devote fifteen hours a day to solid work.'

Edison's first patent, No. 90,646 for an automatic vote recorder, was registered in 1868, his twenty-first year. Technically, the recorder was a success, but politicians rejected the idea on the grounds that it interfered with the filibuster system. Edison's first real money came from his improvements to stock-ticking machines. The 1876 patent for Edison's electric pen, which pierced holes in waxed paper for the purpose of making multiple copies, was later sold to Mr A.B. Dick and became the precursor of the mimeograph process. When Edison went to Western Union in 1874 with his quadriplex system, which enabled operators to transmit more than one message at a time over the telegraph wire, he was paid $10,000. It was on just such a system that Alexander Graham Bell first concentrated his efforts.

Most of Edison's inventions were improvements on nascent technologies – his own as well as other people's. Such was the case with the telephone, when Edison focused the voice through a microphone onto a carbon button, vastly improving the articulation. When Western Union started a telephone company using Edison's transmitter, Alexander Bell and his partners sued and won for patent infringement.

It wasn't long before people who mattered began to take an interest in Edison's work. Investors jostled to back his research and were eager to pay for the rights to patents. In 1876, Edison opened his 'invention factory' in Menlo Park, New Jersey, financed by the sale of the quadriplex telegraph system. His new complex was a combination of industry and research that foreshadowed today's R&D facility. To compensate for the mathematical and engineering expertise he lacked, Edison surrounded himself with specialists, but his relationship with these talented men was complex. Two of his trusted male secretaries failed to run the course: one committed suicide and another suffered a severe nervous breakdown. It was not unusual for Edison to summon his lieutenants in the middle of the night: those men were apt to be away from home for a few days while the team worked out a stubborn problem. Edison usually paid small salaries, intimating that riches would follow successful invention. When cash flow was tight, as it often was, pay cheques could be delayed for months. Sometimes debts weren't paid at all.

Edison was not one to join forces with other inventors and was not above taking full credit for invention when credit should have been shared with others inside and outside his factory. Englishman William Preece, engineer in chief for the Royal Post Office, referred to Edison as the 'professor of duplicity' and described him as a 'young man with a vacuum where his conscience ought to be'. From his early days, Thomas Edison had been a hustler, making his own way in the world, and that was not a habit he was about to break.

On the other side of the coin, Edison's factory was a magical environment for a cadre of talented inventors. At Menlo Park, every conceivable material was available for experiments and, at the head of the ship, was a man whose rich imagination knew no bounds. Make no mistake: Edison was always in charge. He could display a big temper and he was intolerant of laziness, but creativity was encouraged and there was a lot of room for new ideas.

They also had fun at the lab. At midnight, when the factory was a hive of activity, no matter what the weather, someone was despatched to the farm a mile down the road to bring back a hot meal. After the midnight supper, there would be singing and dancing and storytelling, a cigar or two. Supper over, the men – members of what was sometimes called the 'Menlo Park Insomnia Squad' – often returned to their work. It was claimed in the factory that when Edison said he didn't believe in wasting time on sleep, he wasn't talking about himself but about the men who worked for him. One of the tales that Edison spun was that he slept only a few hours a day, if that, throughout his life. In truth, after driving himself and those around him to the end of endurance, Edison would sleep for hours, even days. His ability to nap wherever he was – on a shelf, under his desk, or curled around his machines – was his hallmark, and those naps amounted to more time than he sometimes admitted. He certainly didn't waste much time getting ready for bed: Edison, who cared little about his appearance in general, believed that changing one's clothes 'changed the body chemistry', resulting in insomnia. Edison also attributed his stamina to the fact that 'he did not wear clothes that pinched the blood vessels'.

Henry Ford, one of Edison's closest friends, said that Edison was 'the world's greatest inventor and the world's worst businessman', although he was diligent about filing patents and caveats to cover his tracks. Edison, who hated confrontation, was a poor decision-maker. Sometimes he used investors' money to stock his laboratory; deadlines fell by the wayside and promises went unfulfilled. Edison gathered around him men who were smart, but not too smart, self-starters who were not too independent.

Because Edison rarely went home to sleep, eat, or refresh himself, his family saw very little of him. Even though he worked less than three minutes from his house,

Edison was often absent from home for days at a time. He was not a sentimental man, and his attitude towards women was Victorian. On Christmas Day 1871, he married one of his employees, sixteen-year-old Mary Stillwell, and, a few weeks later, wrote, disappointedly, that 'My Wife Dearly Beloved Cannot invent worth a Damn!!'

Three children later, Mary broke down emotionally, a victim of neglect. At one point, she was consuming a pound of chocolates a day and her weight soared to more than two hundred pounds. The three children from this marriage, the first two of whom were nicknamed 'Dot' and 'Dash', were estranged from their father after Mary's death at the age of twenty-nine in 1884. Plaintive letters from Edison's son Will, away at boarding school, were returned with grammatical corrections and a short note: 'Wm – see marks – your spelling makes me faint. TAE.' When Dorothy Edison, who for a short period was very close to her father, sailed for Europe and a life abroad, the famous inventor declined to come downstairs to say goodbye.

Edison's successful improvements on the telephone and the development of an embossing telegraph, whereby telegraphic messages were imprinted on paper, led Edison to perhaps his most significant and unique invention in 1877, the 'sound recorder' or phonograph. Ironically, in spite of profound deafness resulting from a childhood disease and an overall lack of taste in music and art, the phonograph was Edison's 'prize'. In order to 'hear' the music, Edison would bite the wooden case, absorbing sound through the vibrations – deep tooth marks were clearly visible on the machines in his laboratory and on the grand piano in his house.

When Edison did create a working model of the phonograph and demonstrated it to the editor of *Scientific American*, people crowded the office, in awe of a device that could record and play back the human voice. When the editor turned the crank, words emanated magically from the machine. 'Good morning,' it said, 'What do you think of the phonograph?' During one demonstration, women in the audience fainted at the wonder of it all. It was the phonograph that earned Edison the title 'The Wizard of Menlo Park'.

Typically, Edison made great claims for his inventions in advance of a working prototype, creating demand way ahead of supply. The phonograph was no exception. Introduced prematurely, with its crude representation of the voice, the phonograph's popularity soon subsided, only to be resurrected in full force a decade later.

The phonograph gave Edison the idea for a talking doll, and much of his factory and a large portion of his capital was turned over to that project. A miniature phonograph cylinder was implanted in two-foot-high tin dolls, each of which 'sang' a nursery rhyme. The project was dropped when the mechanism consistently failed to work. A Mr J Walter Fewkes shared Edison's inspired vision of the wide variety of potential uses for the phonograph technology when he carried a hundred-pound machine into the Maine woods and successfully recorded for the first time in history the tale of 'Po-dump, Pook-in-Squiss, Black Cat and the Toad Woman' from the Passamaquoddy Indians.

When Edison claimed to have found the means to provide cheap electric light, including the most powerful generator to date and the components of a modern power distribution system, a group of investors agreed to capitalise the Edison Electric Light Company, but claims of success were once again premature. Demonstrations led to scepticism and anxiety, and no one was surprised when Edison's pledge to illuminate the area around his Menlo Park laboratory with two thousand lights failed to materialise. Edison, a folksy storyteller, conjured up images for the press of the ways in which his inventions would benefit humankind. Often, his grandiose claims upstaged and alienated other inventors, whose serious work was lost amid the glitter of Edison's vaudevillian performance, but he did catch the attention of potential investors. Public declarations also raised the bar for Edison, who was driven to inventive frenzy by the pressure to produce.

'It is easy enough to invent wonderful things and start the newspapers talking,' Edison admitted, 'but the trouble comes when you try to perfect your inventions so as to give them commercial value.'

Light bulbs had been patented before, but they were flawed and impractical for commercial use. By 1879, Edison had produced a high-resistance lamp in a vacuum that could burn for hundreds of hours. Even though troubles continued to plague the project, at the age of thirty-three Edison was worth $500,000 and had become a world-famous inventor. The display of incandescent lamps that eventually lit up the Christmas sky at Menlo Park in 1879 brought people flocking to his door. In January 1880, he filed a patent for the incandescent light bulb, claiming that he would make electricity 'so cheap that only the rich will burn candles'.

Edison suffered frequent bouts of exhaustion and intestinal difficulties, not surprising when considering that he was often under enormous stress, smoked twenty cheap cigars a day and chewed tobacco. As for his deafness, Edison always claimed it as an asset because it provided quiet time for thinking. 'My nerves are intact,' he maintained, 'because I don't hear well.' He also claimed in an interview that he 'did not allow himself to worry,' maintaining that the best way to curtail worry was to keep busy.

In a 1917 interview, Edison shared his views on diet. 'I rarely eat more than six ounces of food at a time,' he said, 'and I boil everything except the water. I eat no lettuce, celery or other raw things in order to guard against bacterial invasion. Chemicals are nowhere near as dangerous as bacteria. I'm loaded with phagocytes,' he continued in his down-home way, 'those friendly little chaps that fight your battle in the blood against disease. I don't want to make their task any harder than it already is.'

Thomas Edison was thirty-nine years old when he married for a second time in 1886, taking as his bride twenty-year-old Mina Miller from Akron, Ohio, the daughter of a wealthy inventor. Edison tapped his proposal on the back of her hand in Morse Code, and she tapped back her answer. 'yes'. Mina, who was only eight years older than her stepdaughter, Marion, was cultured and well educated and made of sterner stuff than her predecessor, Mary. Mina made the most of her forty-six years of marriage to the eccentric inventor, but her loneliness came through when she wrote, 'Everybody is good to me, but my heart is so pinched.'

At the beginning of 1889, Edison was forty-two, a world-renowned inventor, with an attractive and intelligent wife, a large factory, and a million dollars coming his way in one year alone. In spite of ongoing legal battles over patents and rights, the public saw Edison as the inventor of the electric light as well as the phonograph, and he was heralded in America and abroad as royalty.

In fact capitalisation of the Edison Electric Light Company came none too soon, as Edison owed vast sums of money to virtually everyone. A lot of his newfound wealth had gone to finance a vast new laboratory in Orange, New Jersey, designed to mass-produce Edison's inventions. The two brick buildings featured a private office with forty-foot walls, cathedral ceilings, oriental rugs, potted palms and galleried rows of bookshelves. Mina and the children were ensconced nearby in a luxurious house, Glenmont, and the Edison home in Sarasota, Florida – Seminole Lodge – provided a beautiful winter retreat.

The stubbornness that kept Edison on track as an inventor sometimes translated into blind resistance to new trends. The dispute over alternating current was a case in point. Edison discarded the concept of alternating current out of hand, and the lighting industry passed him by. His talented employee, the gifted inventor Nikola Tesla (see page 151), failed to convince Edison that AC current, which could be stepped up to high voltages using transformers, was the way of the future. Tesla finally defected, with his single and polyphase motors and transformers, to Edison's rival, George Westinghouse. One by one, the Edison Company customers switched from direct to alternating current, which was faster and cheaper. Edison, ignoring the pleas of his agents in the field, stood firmly in favour of direct current.

To demonstrate how dangerous AC current could be, Edison the showman staged public electrocutions of hapless dogs, horses and cows. One lucky dog survived 1400 volts and trotted off into the factory bearing his new nickname, Ajax. The coup de grâce was the dramatic electrocution of Topsy, a bad-tempered elephant, at Coney Island's Luna Park. Edison's experiments led to the introduction of electrocution as a means of execution, a process that Edison encouraged and coined 'Westinghousing'.

To generate the enormous capital required to develop the electrical industry, a merger between Edison General Electric and the Thomson-Houston Company was effected and a new entity, General Electric, was formed in 1892. Disgruntled that his name was no longer on the masthead, Edison attended only one meeting as a Director. In fact, according to his secretary Alfred Tate, 'something died in Edison's heart when his name was removed'.

Leaving the electrical industry behind in mid-life, Edison, a millionaire in his own right, purchased twelve thousand acres in New Jersey and set up a mining operation using his electromagnetic iron ore separator. This time he had no partners or investors. As he put it, he was now able to 'paddle his own canoe'. As the debate over AC versus DC current accelerated, Edison buried himself increasingly in his work at the mines, leaving the business of electricity to go where it would.

Edison lost thousands of dollars every month in ore production, deficits he financed by selling blocks of his General Electric stock. He had left behind a burgeoning electrical empire for an increasingly ineffective laboratory and a mining operation that was bleeding him dry. When his rock-crushing operation needed rebuilding, Edison sold the remainder of his GE stock well below par and funnelled the cash back into his mines. That stock would have been worth well over $12 million over the course of Edison's lifetime. Edison's characteristically optimistic response was, 'Well, it's all gone, but we had a hell of a good time spending it.'

While the mining project and the phonograph, beleaguered by court cases, were still at the top of his list, Edison turned his attention in the early 1890s to moving pictures, claiming that his kinetograph (for recording pictures) and his kinetoscope (a peephole device used for showing them) would 'do for the eye what his phonograph had done for the ear'. The Black Maria, an awkward edifice covered in tar paper and named after the paddy wagons which it resembled, was built in 1893 and served as the world's first moving picture studio. Boxing matches were among the activities filmed in the Black Maria, with time out for the hour-long changing of film. If the fighters moved out of range, they were shooed back in front of the camera. One of the first efforts featured Edison employee Fred Ott in the numerous phases of a sneeze.

These viewing machines were already outdated when Edison was approached with a new plan. A Washington, DC, estate agent, Thomas Armat, had developed an effective prototype for projecting pictures onto a screen, but the investors needed Edison's name to market the idea. Very quickly, Edison tweaked the design and appropriated what came to be known as the Edison projectoscope. Not for the first time in his life, Edison defaulted on contractual payments to Armat, who sued.

When the radio came into vogue, Edison maintained adamantly that it would 'never amount to anything' and refused to have anything to do with it.

Eventually, Edison tried unsuccessfully to marry the phonograph with the kinetoscope to create moving pictures with sound. In the mid-1890s, Edison's interest in moving pictures waned and his attention was diverted once again, this time to the news of the discovery of the X-ray by Wilhelm Roentgen.

'We could do a lot before others get their second wind,' wrote Edison, his inventor's curiosity piqued at the thought of being the first to isolate a material that could register the rays after they had passed through a body. His fluoroscope met with success, but some months later, Edison experienced some difficulty with his sight, his hair began to fall out and his skin was peeling. Coupled with other, alarming, reports of the effects of X-rays, including the gruesome death from cancer of one of his assistants, Edison ceased his experiments with radioactivity.

In the late 1890s, Edison returned to his ore project, this time building a large cement plant at great expense. His plan was to create moulds for cement houses, an impractical two thousand three hundred moulds for each dwelling. Edison didn't see any reason why the furniture couldn't also be fashioned out of cement, including pianos, phonograph cabinets and refrigerators. Twenty years later, his Portland Cement plant went bankrupt and the investors lost their money.

Edison's last major project involved the alkaline storage battery, a happy combination of two of his favourite things, DC current and chemistry. The battery business made Edison a millionaire in his later years and sustained his lifestyle more than adequately.

In the early 1920s, Edison put a lot of thought into the definition of a perfect employee and, among other things, decided that a man should have at his fingertips a wide range of knowledge coupled with a good memory. Accordingly, he put together an intelligence test of a hundred and fifty questions, ranging from 'What city manufactures laundry machines?' to the ingredients in a martini. In some respects, the intelligence test was born of Edison's lifelong disdain for formal education.

Edison's opinions, which he never hesitated to express, were anything but cerebral. While abroad, for example, he measured the overall efficiency of peoples in different countries by calculating how long it took them to get out of the way when he blew his horn. Anti-Semitic and prejudiced, he was fond of saying that 'rice-eating nations never progress'. When Rachmaninoff came to his recording studio, Edison's comment was, 'Who told you you're a piano player?'

On the home front, Edison's children suffered on the one hand from their father's neglect and overindulgence on the part of their lonely mothers, although Mina left the upbringing of her three to governesses and private schools. She called Edison 'Dearie' and her children were 'Deariettes'. Tom Jr was insecure and his father was quick to let his son know what a disappointment he was. Searching for something meaningful to do, Tom Edison was an easy mark for shysters who persuaded him to lend his famous name to one scheme or another, including a bizarre medical device called the Magno-Electric Vitalizer. Edison was incensed. Tom's brother, Will, attended Yale briefly before serving in the First World War. His business ventures, including a car repair shop and a poultry farm, invariably failed. Edison was heard to say that his son Will 'brought the blush of shame to his cheek'.

Charles, one of Edison's sons from his second marriage, struggled with alcohol, but he was cheerful and compatible and Edison found him to be good company. When Charles wrote asking for a 'big check' or a 'small check', Edison was apt to open his desk drawer and snip a piece of plaid cloth – big checks or small – which he enclosed in the return envelope. In 1927, Charles took over his father's company, pared down the overheads and tried, too late, to enter the radio market.

Edison enjoyed fishing and played Parcheesi, although he changed the rules to suit himself. Winning was everything. Playing billiards one evening after dinner with a colleague, he missed a shot. Over and over again, the ball was returned to its original spot and Edison practised until he could drop it comfortably into the pocket. Having done that, he announced that he no longer felt like playing.

In October 1931, at the age of eighty-four, Thomas Edison, one of the most famous men in the world, died at home. At the time of his death, he held 1093 patents, singly or jointly, more than any other inventor in the history of the United States.

His legacy as a father was less impressive and his will did nothing to ease relations among his six children, three by each of his marriages. His nine thousand shares of Thomas A. Edison, Inc., valued at $85 a share, were divided equally between Charles and Theodore, his sons by Mina Edison. The $1,390,000 in bonds went eighty per cent to Charles and Theodore; a mere five per cent of the remainder was to be divided among the other four children. All property was left to his wife, Mina. Madeleine, Edison's daughter by Mina and the mother of his only grandchildren, sued, and eventually settled with her brother Charles. Madeleine was very accomplished, perhaps the brightest of all Edison's children, and sat on the board of directors of Western Union. The issue for Edison was probably his disapproval of Madeleine's husband John Sloane, a Roman Catholic.

Charles closed his father's laboratory after Edison's death. Named by Roosevelt as assistant secretary of the Navy, Charles also served as the governor of New Jersey in the 1940s and died in 1969, aged seventy-eight.

Charles' brother, Theodore, a man for whom personal wealth was never of importance, formed a research group, Calibron Industries, to provide jobs for his father's employees when the laboratory closed. He was to live the longest of the siblings, dying in 1992 at the age of ninety-four, the holder of eighty patents in his own right.

Four years after Thomas Edison's death, his son, Thomas Jr, died at the age of fifty-nine, alone in a hotel room in Springfield, Massachusetts. Four days later, Mina married her childhood sweetheart, Edward Everett Hughes. In 1937, Edison's son William died of cancer at fifty-eight.

Thomas Edison left a collection of more than four million pages of notes and papers documenting his life of invention.

The Magician
NIKOLA TESLA

Born: July 9, 1856, Smiljan, Croatia
Died: January 7, 1943, New York City

An inventor has so intense a nature, with so much in it of a wild, passionate quality, that, in giving himself to a woman, he would give up everything and so take everything from his chosen field. It is a pity, too. Sometimes we feel so lonely.

Nikola Tesla

THE TELEPHONE RINGS ON THE HEAD-WAITER'S TABLE at New York's Waldorf Astoria. It is Nikola Tesla calling to place his order for dinner; he will eat precisely at eight. The inventor closes his lab, walks to his hotel where he changes into evening dress and makes his way to the restaurant. He is tall — perhaps six foot six — and weighs a hundred and forty pounds. His angular face and high cheekbones, dark hair parted in the middle and swept back, define an elegant personage. Tesla is shown to a table at the back of the room, away from the other diners. Piled neatly to his left are twenty-four white linen napkins; Tesla takes these, one by one, and polishes his cutlery and china. The used napkins mount in a heap by his chair. With the exception of the few occasions when Tesla hosts exquisite dinner parties, he dines alone.

At precisely ten o'clock, Nikola Tesla leaves the Waldorf and returns, either to his room or to his laboratory, where he often works through the night. He typically takes five hours a day for rest, between 5:00 a.m. and 10:00 a.m., two of which are dedicated to actual sleep. Once a year, he sleeps five consecutive hours, 'storing up rest' for the days to come.

At exactly midnight on July 9, 1856, Nikola Tesla was born in Croatia, then part of the Austro-Hungarian Empire. His father was a minister; his mother an inventor of many small things. From a young age, Tesla envisioned grand possibilities, including an underwater mail tube connecting Europe and America, and was able to design and 'build' precise models in his 'mental workshop'. When translated into actual prototypes, these imaginary models were perfect to the smallest measurement.

Following graduation from university in Prague, Tesla went to Budapest to work in the new telephone exchange. Walking through the park one evening at sunset, Tesla conceived the rotating magnetic field and, with it, a vision of the way in which an alternating current system could be activated through dynamos, motors, transformers and other electrical devices. He also worked out a way to operate three or more alternating currents simultaneously, his 'polyphase' system. These ideas were fundamental to the development of large power systems. In contrast, Thomas Edison, committed financially and philosophically as he was to direct current, could only supply power within a one-mile radius of the generating station.

From Budapest, Tesla moved to Paris where he worked for the French-owned Continental Edison Company, which, under the Edison patents, installed early lighting systems in Europe. When an explosion rocked the opening of a railroad station in Alsace, with Emperor William I in attendance, the Germans cancelled the power installation. Tesla was promised a handsome settlement if he could repair the situation. His task accomplished, Tesla returned to Paris, where Continental Edison declined to honour their pledge. Angered by the breach of promise and frustrated by their lack of interest in his proposals for an alternating current system, Tesla resigned. Continental Edison not only lost a brilliant engineer, but they also lost their chance to capitalise on an invention worth millions.

Armed with a letter of introduction to Thomas Edison in America, twenty-eight-year-old Tesla sold what he could to raise money for the journey across the Atlantic. On his way to the railway station in Paris, his few belongings were stolen and Tesla disembarked at Battery Park in New York in 1884 with only 4 cents in his pocket and the clothes he stood up in.

In his letter of introduction to Edison, Charles Batchelor wrote this famous line to his friend in New Jersey: 'I know two great men, and you are one of them; the other is this young man, Nikola Tesla.' Each was a visionary, but Tesla and Edison were as different as two men could be, not only in method, but in temperament. Where Tesla was convinced that the alternating current system was the way of the future, Edison passionately defended the direct current system as the only way. Tesla worked out problems in his head and came up with an almost perfect model; Edison went through hundreds and sometimes thousands of iterations in his trial-and-error method of discovery.

In spite of these differences, Edison appreciated the young Serb's potential and hired him to work in the laboratory. When Tesla suggested ways in which Edison's dynamos could be altered to improve efficiency, Edison offered him $50,000 to produce a design. The task completed, Tesla returned to Edison for his bonus. 'Tesla,' the American inventor replied, 'you just don't understand our American humour.' Tesla never received a penny of the promised sum. In the spring of 1885, less than a year after their meeting, Tesla resigned. Edison was to say at the end of his life that the biggest mistake he ever made was losing Nikola Tesla.

With very little money and no capital to fund the development of the ideas blossoming in his head, Nikola Tesla accepted an offer from backers to set up a company, the purpose of which was the development of the arc light for use in factories and street lamps. Tesla's theories about alternating current and its possibilities were of no interest to them. Tesla was paid a small salary with the added incentive of stock in the company. Patents in hand and production underway, Tesla discovered that the stock certificate issued in his name was worthless. Once again, he had been cheated.

Tesla referred to the time that followed as his 'year of terrible heartaches'. From the spring of 1886 to the spring of 1887, he lived in abject poverty, taking jobs as a day labourer and surviving as best he could. In the midst of hopelessness, a works foreman, impressed by Tesla's ideas, introduced him to Mr Brown of Western Union. Brown and a partner financed the Tesla Electric Company in 1887 and Tesla soon had working units employing the polyphase alternating current system. Edison's current was limited to 220 volts; Tesla could now transmit currents of many thousands of volts. The War of the Currents was underway.

People began to take notice when a series of fundamental patents was issued to Tesla in 1888. An invitation to address the prestigious American Institute of Electrical

Engineers followed. Tesla's famous lecture, which outlined the theory of the practical application of alternating current to power engineering, won him instant recognition and, overnight, Nikola Tesla became a hero.

One observer equates Tesla's discoveries with a 'ten pound diamond' – the value is indisputable but no one knows quite what to do with it. The next step called for a white knight with vision and capital, and that figure appeared for Tesla in the form of George Westinghouse. A few weeks after the Institute lecture, Westinghouse made an appointment to meet at Tesla's New York laboratory.

Westinghouse wasted no time. The two men liked and respected one another and soon struck a deal whereby Westinghouse would pay Tesla $1 million for his patents, numbering around forty. In addition, fully aware of the intrinsic value of this great discovery, Westinghouse agreed to pay a royalty of $1 for every horsepower of energy produced by the system. A handshake sealed the agreement. Tesla repaid his backers at Tesla Electric $500,000: the remaining money made the thirty-two-year-old inventor a rich man.

Westinghouse had struck a handsome deal, but he had miscalculated and soon found himself overextended. An economic depression changed the climate in America and the merger of two of his largest competitors, Thomson-Houston and Edison General Electric, put a new edge on competition. Westinghouse had a clear vision of growth based on the Tesla patents, but he did not have sufficient capital to move forward. His financial backers demanded reorganisation, including the cancellation of Tesla's royalty agreement.

George Westinghouse was a fair and honest man caught in an impossible situation. The journey to Tesla's laboratory at 33 South Fifth Avenue in New York must have been a painful one for the industrialist. Westinghouse came right to the point. There was a choice, he explained. Tesla could demand his rights and insist on royalties, in which case the Westinghouse Company would surely fail. If Tesla was willing to release Westinghouse from the contract, the company would survive to develop Tesla's ideas. It was as simple as that.

Nikola Tesla thought a minute, then gave his decision. 'The benefits that will come to civilisation from my polyphase system,' he said, 'mean more to me than the money involved.' Tesla tore up his contract and, in so doing, waived his rights to millions, perhaps billions, of dollars and a life of financial ease. With the remainder of the initial million-dollar payment in hand, Tesla must have felt like a rich man, but invention is an expensive business and development capital soon dried up. Because Tesla's fatal flaw lay in his resistance to collaboration, the termination of his contract with George Westinghouse – a perceptive man whom he liked and respected – marked a major turning point in Tesla's career.

What Tesla did have on his side was a wealth of ideas and unshakeable confidence in his own abilities, but for all his genius, he had no business sense. Inventions poured forth but were never patented. Opportunities to translate his work into lucrative projects were sacrificed in order to push ahead with his experiments. Tesla was convinced that he would live to be a hundred and twenty-five; there was time enough to bring his ideas to fruition.

By the early 1890s Tesla, elegant and mysterious, was very much a part of New York's intellectual elite. His dinner parties, held at the Waldorf Astoria, were among the choicest of invitations. Warmed by superb food and wines, Tesla and his guests, all in formal attire, walked the short distance to the laboratory where Tesla turned from host into magician. Objects whirled through space and glowed eerily in the dim light. Sparks and sheets of flame surrounded the onlookers. As his *pièce de résistance*, Tesla stood on a platform while thousands of volts of electricity passed through his body and lit a lamp he held in his hand.

While Edison was electrocuting dogs and cats to demonstrate the dangers of alternating current, Nikola Tesla had discovered that while low-frequency alternating currents were painful and damaging to the human body, very high-frequency currents were harmless. His exploration of the range of electrical vibration between electrical current and light waves marked the end of Edison's claims. The high-frequency current transformers that Tesla developed to produce high voltages became known as Tesla Coils.

You can imagine that Tesla's neighbours on New York's Houston Street were puzzled. Odd lights, flashes of sparks and flame, the gaunt figure of Tesla coming and going at all hours, must have excited speculation. Then came the earthquake.

Tesla was fascinated by the vibrations caused by his mechanical oscillator and wondered about the positive effects these might have on the body. Attaching his small oscillator to the iron pillar in the middle of his laboratory, Tesla built up an ever-increasing frequency of vibrations. Chairs and desks in the nearby police station rattled across the floor; neighbourhood windows shattered and plaster fell from ceilings and walls. The earth shook. Gas and water pipes split apart. Dashing to the laboratory, the police arrived just in time to see Tesla smash a small box attached to the pillar in his laboratory. A heavy silence ensued. All was still. 'Gentlemen,' he said, turning to his audience, 'you are just a little too late to view my experiment.'

A decisive victory against Edison came when Tesla's technology was used to light the 1893 World's Fair in Chicago. In 1896, Tesla's boyhood dream came true when his polyphase alternating current system was employed to convert the hydraulic power of Niagara Falls into electrical energy, energy that was then transmitted twenty-two miles to light Buffalo, New York. This successful demonstration of Tesla's theories opened the door to an unlimited number of fantastic new inventions, inventions that many people saw as nothing short of magic.

It is at this juncture that Tesla's inability to convert his inventions into commercially viable ventures changed the course of history. Not only was the inventor naïve in matters financial, but he was obsessed with his work and unwilling to prosecute patent infringement. Litigation took too much time from other, more important, things.

On March 13, 1895, Tesla's laboratory burned to the ground and, without insurance, all was lost. Tesla was bankrupt. In desperation, he accepted $100,000 from the head of the Morgan Group, but when a further alliance with the Morgan empire was proposed, an alliance that would end all Tesla's financial difficulties, Tesla turned them down. He believed firmly that his inventions would earn him money over time. Time had also shown that Tesla did not work well under the restrictions imposed by a bureaucracy.

Morgan's $100,000 bought Tesla a new laboratory and he set to work on wireless transmission, establishing many of the basic principles of radio technology. Soon he was broke again, sustained here and there by personal donations from the likes of John Jacob Astor and J. Pierpont Morgan. Abandoning New York, Tesla moved to Long Island, where he built a mammoth tower for the development of his wireless technology and the establishment of a world broadcasting station. Armed guards surrounded the facility to prevent spying. When creditors, including the Waldorf Astoria, called in their loans, Tesla was forced to close down his lab.

In 1912, Tesla was awarded a joint Nobel Prize with Thomas Edison. As meaningful as the $20,000 prize would have been, Tesla refused to be associated with his former employer. When the Committee heard that Tesla would not appear on the stage with Edison, the award was rescinded.

The last twenty years of Tesla's life were spent in penury. He subsisted on a small honorarium from the Yugoslav government and a gift here and there. Suffering from obsessive-compulsive disorder, he discarded handkerchiefs after one use, refused to shake hands, and could not eat a meal if someone at the table was wearing pearls. The sight of a peach was too much to bear. Nikola Tesla lived a celibate and isolated life.

Often, after midnight, a lone figure could be seen walking down Fifth Avenue towards St Patrick's Cathedral or, farther, to the New York Public Library. There Nikola Tesla would wait for the soft whistle and the whoosh of wings, a flock of pigeons arriving for their seed. The tender moment over, Tesla would walk back to his hotel and close the door behind him. On the table you might see a number of baskets, a pigeon cooing in her soft nest; perhaps his favourite, a white bird with grey wings, would be there. When Tesla was ill, unable to visit his leathered friends, he would pay a young boy from Western Union to take the seed for him. Unlikely companions for a man obsessed with cleanliness, but the only family Nikola Tesla had.

In 1943, Nikola Tesla died alone and penniless in his hotel room at the New Yorker Hotel. His only possession was the gold Edison Medal awarded to him in 1917. Perhaps because of Tesla's Slavic connections, FBI agents broke down the door and confiscated all his papers and files. They were never seen again. Tesla died holding seven hundred patents and with hundreds and perhaps thousands of ideas stored in his mental file, visions that never saw the light of day.

Where in the world?
LOUIS AIMÉ AUGUSTIN LE PRINCE

Born: August 28, 1842, Metz, France
Died: September 1890, somewhere in France?

For all sad words of tongue and pen, the saddest are these, 'it might have been'.
John Greenleaf Whittier

LE PRINCE'S LIFE IS A SAD STORY. I have to think back to Isambard Kingdom Brunel's admonition to fellow inventors that to focus on that one big dream can actually hold back creative genius. Le Prince did exactly what Brunel advised against, spending all his time and energies on 'one grand scheme'. And he came so close, but not close enough. Le Prince is a footnote in the history books, secondary to dynamic and colourful figures like Edison and the Lumière brothers.

As Thomas Edison was passing through London after a triumphant visit to Paris for the Exhibition in 1889, Louis Aimé Augustin Le Prince was sequestered in his workshop at 160 Woodhouse Lane in Leeds, working on his prototype for the motion picture camera and projector. Three thousand miles away, at Edison's factory in West Orange, New Jersey, William Dickson had taken advantage of his boss's visit to Europe to resurrect their own moving picture experiments.

Le Prince, born in 1842 in Metz, a French town not far from the German border, favoured art over mechanical invention, although he studied optics and chemistry at university. His father, a major in the French Army, prevailed on his good friend Daguerre to give Louis lessons in photography, and it appeared that he would follow a career as a painter and ceramicist.

A handsome and gentle man who stood six foot four in his stocking feet, Le Prince met a young Englishman, John Whitley, while at university and accepted an invitation to visit the north of England. In Leeds, Louis Le Prince was introduced to John's father, Joseph, owner of a successful foundry and an inventor in his own right. On that visit, Le Prince also met John's sister, Elizabeth, an artist who had studied pottery in Sèvres, France. Their love blossomed into marriage and, over time, the union produced six children.

For some time, life was good. Lizzie and Louis opened the Technical School of Applied Art in Leeds and Le Prince perfected a method for firing coloured photographs onto pottery. Among other commissions, he produced two portraits, one of Queen Victoria and another of Prime Minister Gladstone, which were placed in a time capsule in the foundations of Cleopatra's Needle on the Thames Embankment in 1878. An honourable man, brought up in the military tradition, Le Prince returned to Europe in 1870 to fight in the Franco-Prussian War, where he survived the siege of Paris.

In 1881, Le Prince was persuaded by his brother-in-law, John, to move to America, where they held the rights to a new interior design process known as Lincrusta wallpaper. In America, Lizzie became an art teacher and later head of the art department at the New York Institute for the Deaf. Le Prince moved on from his interior design work to become manager of a group producing panoramas, 360° circular representations that produced the sensation of being 'inside' the picture. Like many others, Le Prince had seen the zootrope of Eadweard Muybridge and admired his rapid sequence photography of animals in motion. Both Le Prince and Edison had also seen the work of Étienne Jules Marey and his photographic 'gun', which could shoot up to twelve exposures a minute.

Experienced in the effects of the panorama and full of ideas about ways in which he could animate the panoramic illusion, Le Prince embarked in 1884 on the creation of a camera and projector for the production and display of moving pictures. With his experience as an artist and photographer and knowledge of optics and engineering, coupled with insights into invention through his father-in-law, Le Prince set out on his new course.

Through 1885 and 1886, when not working on the panoramas, Le Prince retreated to the laboratory to work on moving picture models. When funds ran short, he considered approaching Edison for backing in his moving picture project, but was advised strongly against such a move; Edison was not to be trusted. In the end, he turned instead to his father-in-law in Leeds, an enthusiastic supporter, who could supply the infrastructure necessary to carry on with moving picture experiments.

Promising to be away no more than a year, Le Prince left his family in America in 1886 and returned to the north of England. Once in Leeds, he hired woodworker Frederic Mason and engineer J.W. Longley, the inventor of the automatic coin-operated ticket-dispensing machine. Together they set up shop and work began in earnest. Mason described his employer as a 'most generous and considerate person, and, although an inventor, a man of extremely placid disposition, which nothing seemed to ruffle'.

Work on the camera progressed fairly smoothly. The projector was another matter, not least because they were unable to find film sufficiently flexible, heat resistant and transparent. In the early days, paper film was all that was available; celluloid appeared in late 1889, but it was thick and unwieldy at first and bubbled under the heat of the lamps.

Time and again, the family tried to persuade Le Prince to introduce his new machine at various exhibitions; just as often, he declined, obsessed by the fact that it was not yet perfect. There was in the more modest Le Prince none of Edison's showmanship, no trace of inclination to boast about his accomplishments before they were proven entities. He was also preoccupied with keeping his work a secret.

In 1888, US Patent No. 376,247 was granted to Le Prince, mentioning only the sixteen-lens camera that he had produced in 1886. This was an important turning point. The original application mentioned 'one or more lenses', but when the Patent Office claimed interference with an earlier filing, Le Prince's stateside lawyers agreed, without consultation, to the elimination of the single-lens clause. That adjustment to the patent was to have important negative repercussions in the Edison court case many years later and would throw into doubt Le Prince's place in history. In the same year, British and French patents were issued to Le Prince with reference to a projector and a camera with a single lens as well as multiple lenses.

In 1888, Le Prince made what many consider to be the world's first successful attempt to record moving images. The 'film' was taken in the garden at the Whitley home in Leeds. Approaching the heavy single-lens camera mounted on a sturdy base, Le Prince turned the crank and captured on paper film a family group as they walked across the lawn in the lengthening shadows. The film lasted a few seconds. Another short film, depicting traffic and pedestrians crossing a bridge in Leeds, is seminal and the twenty or so frames are accepted as proof that Le Prince did, in fact, perfect and use a single-lens camera.

Promise followed promise and year followed year as Le Prince assured his wife Lizzie that he would 'soon be done' and return to New York. The one year abroad had stretched to three. Finally, encouraged by the availability of a more versatile celluloid film, he advised his wife to secure a place in New York where he could introduce his camera and projector to the public. Lizzie made preparations for her husband's debut.

Ironically, when Lizzie travelled to California in connection with her own work, Governor Leland Stanford of California and his wife invited her to become head of the art department at Stanford University, due to open in October 1891. The salary was very attractive and she tried in vain to persuade her husband to accompany her and the family to the west coast, where he could continue to work on his camera and projector. Governor Stanford had previously supported Eadweard Muybridge in his experiments with the zooscope, so it is entirely probable that he would also have provided critical financial support to Le Prince.

Failing to appreciate his own shortcomings as a businessman and the limitations on his project due to lack of capital, Le Prince refused to consider the offer. He was neither a salesman nor an entrepreneur by nature and missed an important opportunity to link up with a benefactor who could easily have been both. 'Moving pictures will solve our money worries,' he wrote to Lizzie, 'Just you see.' He added, 'The financial returns from my work will dwarf anything we might make from a new art school venture.'

The pressure mounted. With no income to offset major expenditures, bills piled up. Increasingly, reports of technical advances in the motion-picture field filled magazines and newspapers, adding to the sense of urgency. Highly sophisticated in the matter of patents and caveats and never shy about jumping into the arena, Thomas Edison filed his first moving picture patent in 1888, long before he had anything to show for it. On the domestic front, Lizzie was impatient and lonely. Le Prince's father-in-law, who had come on hard times financially, was virtually bankrupt.

In 1890, Le Prince finally declared himself ready to return to New York. In September, just before leaving for America, he travelled to France to visit his brother in Dijon. Their meeting centred around Louis' claims to an inheritance he felt due from his mother's estate. The brothers met, ostensibly they quarrelled, and on September 16, Louis boarded the train in Dijon. He carried with him a small valise containing his patents and papers.

When the train pulled into Paris, Le Prince was nowhere to be found; there was no trace of the gentle inventor. When his workshop was entered a few months later, nothing had been moved. With Le Prince's banker, a man who had no detailed knowledge of his work, in charge of clearing out the laboratory, anything but 'finished' pieces were discarded. Drawings and papers that would have proved invaluable in establishing Le Prince as the true inventor of a single and multi-lens camera and projector were committed to the dustbin.

For years, while she continued her work at the School for the Deaf and raised her family, Lizzie Le Prince maintained hope that her husband would reappear. En route to Panama to visit her son, her ship collided with another just outside New York harbour and went down. Lizzie was thrown overboard and managed to stay afloat, but a falling spar hit her chest, resulting in the removal of her right breast.

Lizzie Le Prince doggedly put forward her claims that her husband had been the true inventor of the moving picture camera, standing helplessly by as Edison garnered all the credit. Edison was a powerful foe and until her husband was declared legally dead, a process that took seven years, Lizzie was unable to press ahead legally.

Le Prince's son, Adolphe, who had worked with his father on the camera and projector from 1887 to 1889, testified at the trial in the Patent Wars over the rights to the moving picture camera and projector. Without the mention of a single-lens camera in his father's US patent and with no working drawings to present, Adolphe had little to counter Edison's claims. When Edison won in the courts, Adolphe Le Prince was crushed. Not long after Edison's triumph, Adolphe was found dead from a 'hunting accident', a gun by his side.

Lizzie and Louis' son Joseph became a colonel in the US Army and settled in the south, where he worked to eradicate yellow fever. Lizzie died in 1926, thirty-six years after her husband's disappearance. In 1930, their daughter Marie carried Le Prince's single-lens camera to England for the unveiling of a plaque in Leeds honouring her father.

What happened to Louis Augustin Le Prince? Lizzie fervently believed there had been foul play, most probably at the instigation of Thomas Edison. Others were convinced that the pressures of mounting debt and Le Prince's obsession with the perfection of his projector had driven him to suicide.

There is a third possibility. Taking into account Le Prince's background as a military man, some believe that he joined the French Foreign Legion, finding honour in service and a new life away from the eye of the camera.

The magic box
AUGUSTE AND LOUIS LUMIÈRE

Auguste Lumière
Born: October 19, 1862, Besançon, France
Died: April 10, 1954, Lyon, France

Louis Lumière
Born: October 5, 1864, Besançon, France
Died: June 6, 1948, Bandol, France

*The cinema is an invention
without any commercial future.*

Louis Lumière

PEOPLE WALKING DOWN THE BOULEVARD DES CAPUCINES in Paris three days after Christmas, 1895, would have noticed a man standing at the entrance to the Grand Café. In response to his invitation to 'see the show', thirty-five people paid the one franc admission, accepted a programme and descended the wooden stairs into the basement. A white sheet hung on one wall. A piano player tinkled the keys as people found their seats.

Suddenly the lights dimmed, with the exception of one behind the man and his wooden box at the back of the room. The first image, a factory entrance, filled the screen. Nothing much interesting here, but as the operator started to crank the handle, the factory gates opened. Workers on their way home walked across the screen and disappeared. About a minute later, the film ended.

It didn't take long for the operator to start the next film, *The Arrival of a Train at la Ciotat Station*. The great locomotive roared out of the distance towards the audience while nervous viewers, seemingly in its path, shifted in their seats or stood up, ready to flee. In *A Sprinkler Sprinkled* the Lumière Brothers played a trick on their father's gardener. Stepping on his hose to cut off the flow of water, the brothers watched as the gardener picked up the nozzle in puzzlement. When Louis stepped off the hose, the gardener received a squirt in the face. The audience roared with laughter. Outside, on the street, they argued about the miracle they had just seen. How was it done?

Word of this magical show spread, and within a few weeks, lines of eager viewers extended down the Boulevard waiting for a chance to see the Cinematographe in action. The daily take at the box office ran as high as 2000 francs, a shock to the café owner who had refused an offer of 20% of the box office receipts in favour of a guaranteed 30 francs a day. The Lumière Brothers had achieved what other inventors were fervently hoping to call their own – the art of the moving picture.

Auguste and Louis were born in Besançon, France to portrait painter Antoine Lumière and his wife. Louis, suffering from chronic headaches and insomnia, left his prestigious technical school and, by 1881, at the age of seventeen, had invented the 'Etiquette Bleue', a much-improved, highly sensitive photographic plate. A decade after the Lumière factory opened in 1883, it ranked as the second largest producer of photographic products in the world, behind Eastman Kodak. The Lumières were manufacturing more than fifteen million photographic plates a year.

Towards the end of 1894, Antoine Lumière visited Paris where he attended a showing of Edison's new peep-show kinetoscope. Edison was, he reported back to his sons, selling his peep-show film at 'insane' prices. Antoine suggested that his sons get to work on a moving picture system that improved on Edison's model, which was accessible to only one viewer at a time. 'Get the image out of the box,' he advised, seeing the advantage of having not just one, but many people at a time enjoying a film.

The brothers got to work and by February 1895, with remarkable speed, had patented a camera/projector. By March, the Lumière brothers had made their first film, workers leaving a factory, using a camera/projector weighing only sixteen pounds – a camera far lighter and more portable than Edison's. The shutter on the Lumière camera opened and closed sixteen times a second, creating a flickering effect, hence the 'flicks'. Thirty-five millimetre film, a size that became standard in the industry, was moved forward by the classic sprocket and pin system used in modern cameras, a concept adopted from the sewing machine.

The Lumières' rapid success must have been a bitter pill for fellow countryman Louis Le Prince, who, eight years earlier, had successfully recorded moving images when he produced the *Roundhay Garden Scene* in Leeds (see page 159). In spite of that success, Le Prince failed to bring his camera/projector to the public's attention. The situation would have been even more frustrating for Léon Bouly, a young French engineer who filed a patent for the 'Cinématographe Bouly' in 1893. When Bouly was unable to pay the annual patent fees, the Lumières picked up the expired patent, including the name, and the young French inventor was relegated to a footnote in history.

What the Lumière family did have was a powerful combination of assets in the world of invention. Technically competent artists and true visionaries, the Lumières were also successful businessmen and gifted promoters, with the capital to support their ideas and a healthy sense of urgency. For the first showings of their new invention, the Lumières chose the Société d'Encouragement à l'Industrie Nationale in Paris, where the show was interrupted frequently by applause, and the Congress of the French Photographic Society. Sharing their invention early on with members of the establishment was an important strategic decision. As a creative and confident team, the Lumières were spared the isolated vigil of the lone inventor. They were not going to be intimidated by the Edisons of the world.

The Lumières also made a strategic decision as to the best way to realise a profit. Rather than respond to the increasing clamour for cinematographes, the Lumières decided to restrict the production of machines by others and amass money instead from the making and showing of films, the rights to which were tightly controlled. By 1897, the Lumière catalogue contained more than seven hundred titles. These short films were essentially animated photographs, with no story line or character development, but they were revolutionary in their day.

The time was ripe and the Lumières pressed ahead with their advantage. Armed with a camera and projector light enough to be carried by hand, the Lumières hired cameramen who would travel the world, making the first documentaries. Suddenly, audiences in Paris or London could see people walking along the streets in Moscow or Jerusalem. One gifted film-maker took the first recorded action shot from his perch aboard a gondola on the canals of Venice.

The 'Wonderful Living Pictures' made on the Lumières' new camera/projector were shown at Britain's Royal Polytechnic Institution on Regent Street in February, 1896, where audiences were treated to the sounds of a train, the splash of breaking waves, or the clip clop of horses' feet sounding from behind the screen. One account reported that a nurse had to be brought in to see to the ladies who had fainted during the arrival of the train. A single projectionist, dressed in tailcoat and top hat, might work at a number of music halls in the course of an evening, travelling between venues in a cab while frantically rewinding his films.

So stunning was the effect of this new technology that when the films were shown in Russia, the audience, believing they were experiencing black magic, threatened the operator. That night, the theatre and all the equipment in it were set on fire. The Lumière films had their New York Debut at Keith's Union Square Theatre. On June 29, 1896, an attempt to market Lumière equipment in America was blocked when Customs accused them of illegal importation. The McKinley administration, perhaps as a gesture towards their own resident genius Edison, essentially forced the Lumières to withdraw from America less than a year after the first screening.

The last Lumière film was projected onto a large screen, ninety-nine by nine feet, at the 1900 Paris Exhibition, following which the brothers sold all rights to their camera to Charles Pathé. Pathé saw the budding field somewhat differently: 'The cinema will be the theatre,' he pronounced, 'the newspaper, and the school of tomorrow.'

As creative and stimulating as the exercise had been, the Lumières were not movie makers or showmen at heart. From a practical point of view, the cinematographe, like the cotton gin and the telegraph, was a juggernaut that defied monopoly, too big an idea to constrain with licenses and lawsuits.

The cinematographe was hardly the end of the Lumière story. Early in the 1900s, they developed a highly successful colour process known as the Autochrome Lumière. Louis went on to create a photographic method of measurement and Auguste worked in the medical field. Far from regretting their exit from the movie business, Louis was heard to say late in life that if he had known the future of the cinema, he would never have invented it.

In 1995, forty of the world's most famous directors produced a retrospective using the original camera from 1895. It worked perfectly.

Doctor Beep Beep
WILLIAM BRADFORD SHOCKLEY

Born: February 13, 1910, London, England
Died: August 12, 1989, San Francisco, California

If you take a bale of hay and tie it to the tail of a mule and then strike a match and set the bale of hay on fire, and if you then compare the energy expended shortly thereafter by the mule with the energy expended by yourself in the striking of the match, you will understand the concept of amplification.

William Shockley

THERE WAS EVERY REASON TO BELIEVE THAT William Shockley would lead a charmed life. By the age of twenty-two, Shockley was not without quirks, but he was indisputably brilliant and eager to succeed. There were, however, some fatal flaws: hubris coupled with a highly competitive nature, paranoia and a formidable temper.

In 1908, Shockley's father had joined a mining firm and moved his wife, May, to London, where, in February 1910, their only child, 'Billy', was born. When looking for accommodation, May was known to visit a potential flat more than ten times before deciding. Little Billy was home-schooled until he was eight, but when his temper tantrums accelerated out of control, William and May sent him to a military academy. When Billy was fifteen, his father died of a stroke.

College at CalTech was followed by four years at MIT, where Shockley adopted a signature that he used until the end of his life, W=Shockley or W=S. While at MIT,

Shockley played elaborate practical jokes, took up rock climbing and expanded his skills as an amateur magician. He raised ant colonies in large glass containers and spent hours training the ants to follow circuitous routes in their search for food. On a later visit to South Africa, most of his spare time was spent evaluating the trainability of local ants.

In the early part of the twentieth century, AT&T, working under Alexander Bell's original patent, was looking for ways to replace the vacuum tube and better amplify sound over long-distance wires. By 1925, Bell Laboratories had been formed, a cadre of physicists and engineers working exclusively on new developments for AT&T. Married with a child, Bill Shockley joined that team in 1936 to study devices that controlled the flow of electricity. An unconventional thinker with the ability to go to the heart of a physical or mathematical problem and devise unique solutions, William Shockley was the man for the job. For recreation, Shockley often scaled the stone façade of the Bell Labs' cafeteria.

While engaged at Bell Labs, Shockley responded to a call from Washington to tackle a number of problems facing the military as they entered the Second World War. Applying statistical methods to military strategies, Shockley and a band of scientists greatly increased military efficiency, in particular in the field of anti-submarine warfare. By 1945, Shockley had risen to become expert adviser to the Secretary of War.

Work may have been stimulating, but the Shockley marriage was disintegrating and Bill was depressed. He wrote a note to his wife, Jean, telling her that he felt 'less well suited than most to carry on with life', then loaded one of the six chambers of his revolver, pressed the muzzle against his temple and pulled the trigger. Surviving his flirtation with Russian roulette, he wrote a second note to his wife apologising for his failure to 'solve their problems'. Both notes were found in Shockley's safe after his death forty years later.

For five weeks beginning in November 1947, a period Shockley referred to as the 'Magic Month', Walter Brattain and John Bardeen, under Shockley's part-time leadership, focused on the invention of the point-contact transistor at Bell Labs. The first prototype, no larger than the end of a shoelace, involved a spring made from a paperclip and two sections of gold foil placed 0.04 cm apart. On Christmas Eve, they successfully demonstrated the device's ability to amplify speech.

Although Brattain and Bardeen had incorporated many of Shockley's theories, credit for the invention of the point contact transistor really belonged to them. Shockley believed that his should be the only name on the patent and his angry statement to that effect further distanced him from his team. In the end, Bell Labs' patent attorneys determined that only Brattain and Bardeen's names should appear on that patent; they applied for a second patent for the junction transistor in Shockley's name alone. Both patents were filed in 1948. With the patents filed, Bell Labs called a press conference. May Shockley was in the audience when her son, speaking for the group, described the point-contact transistor. With a couple of exceptions, the press, much to the scientists' surprise, was singularly unimpressed. The invention of the transistor earned only four column inches in the *New York Times*' radio section.

Fuelled by frustration at being left out of the initial discovery, Shockley pulled away from the group and went to work on a junction transistor, a more robust amplifying device and the precursor of every transistor that powers the electronic age. The first junction transistor was built in 1951; that, and the work that preceded it at Bell Labs, ushered in the electronic revolution. In 1953, *Fortune* magazine referred to Shockley's invention as the 'pea-sized time bomb.' In June 1948, Bell applied for a patent for the junction transistor, which promised to solve the problems of the old vacuum tubes and which could, potentially, be manufactured in a microscopic size.

As tensions increased between Shockley and his original team members, Brattain and Bardeen did what they could to patch things up. To the letter that Brattain wrote in an effort to make peace, Shockley attached a handwritten note to his file copy. 'I am overwhelmed,' he wrote, 'by an irresistible temptation to do my climbing by moonlight and unroped.' The three men later shared a Nobel Prize for Physics for the invention of the transistor.

Jean Shockley was recovering from uterine cancer in 1953 when Bill announced that he would be leaving her and their three children. A cold and unaffectionate husband and father who declared that he had never wanted children in the first place, it didn't take him long to meet and marry a psychiatric nurse, Emmy, with whom he spent the rest of his life. When asked why he had chosen her, he answered, 'Because she understands people better than anyone I know' – an answer of which she was very proud. When he announced his divorce to his mother, she wrote back warning him about the perils of the male menopause.

Along with a new wife, Bill Shockley decided on a new career, one that would potentially enable him to earn substantial money. As brilliant as he was as a physicist, his abrasive management style got in the way of promotion at Bell Labs and, in 1953,

Shockley resigned. The new vehicle, Shockley Semiconductor, was a laboratory funded by Arnold Beckman of Beckman Industries, where a team of brilliant scientists, hand-picked by Shockley, would develop and produce transistors and semiconductor devices. Shockley Semiconductor opened its doors in 1955.

Initially, Shockley's intention was to develop silicon transistors, but within a short time, he had changed course, focusing instead on the Shockley Diode. His research team was unanimous in its disagreement with this decision, but Shockley, stubborn as always, remained adamant. His abrasiveness and paranoia continued to drive people away. When a secretary cut herself opening a door, Shockley was convinced that she had been sabotaged and ordered lie detector tests for all staff. The culprit turned out to be a broken drawing pin.

Working relationships deteriorated as Shockley yelled insults and demeaned those around him. Within a year, Shockley Semiconductor, with no product to speak of, was failing and his team members were mutinous. Two of the men he had hired, Gordon Moore and Robert Noyce, had the entrepreneurial sense that evaded Shockley and continued to work quietly on the silicon transistor, convinced that this was the way to go. In 1957 the 'Traitorous Eight' left to form Fairchild Semiconductor, where they continued their research. Noyce and Moore eventually established their own company and called it Intel. Thirty years later, Moore alone was worth $8.8 billion.

It wasn't long before Shockley, whose interpersonal skills went from bad to worse, fell out with his backer, Beckman. When Fairchild and Texas Instruments, among others, introduced the first integrated circuits, Shockley's work on the diode became superfluous. In 1969, as the Silicon Valley companies that sprouted up around him were making fortunes, Shockley Semiconductors faded off the scene, having never made a profit.

In 1961, Shockley, a man who prized physical fitness, was driving in the fog south of San Francisco when a car swerved into his lane. Shockley, who hit the windscreen, was in hospital for a month, his wife for six. Increasingly antisocial and insecure, perhaps permanently impaired by the blow to his head, Bill Shockley now turned his attention to genetics, specifically the theory that intelligence was inherited and racially based. A great admirer of the intellect, Shockley worried that the unfit would survive and multiply at a greater rate than those of superior intelligence. During a television appearance at which Shockley expressed these views the audience erupted in anger, draping him in a swastika-emblazoned banner. Invited to speak about the development of the transistor at a large university in South Africa, Shockley took his opportunity at the podium to speak about the relative merits of various ethnic groups, much to the embarrassment of everyone present.

In an effort to promote superior traits, Shockley announced that he had contributed to a sperm bank, the Repository for Germinal Choice, which initially solicited donations only from Nobel Laureates. A headline in the Atlanta newspaper announced, 'Designer Genes by Shockley'. In a *Playboy* interview, he explained that his first wife, Jean, was his inferior intellectually, and referred to his own children as a 'significant regression' relative to himself. Yet his daughter graduated from Radcliffe and his eldest son from Stanford (although his other son, did not earn a degree).

Few friends weathered Shockley's insults and churlish behaviour; eventually, most of them jumped ship. It was not unusual for Shockley to arbitrarily leave the dinner table, no matter how august the company, saying that he would be glad to continue the conversation when others were better prepared. He brooked no contradictions and had recording devices installed on all his telephones, the intermittent beeping of which earned him the moniker 'Doctor Beep Beep'. Every telephone conversation for the rest of his life, no matter how pedestrian, was taped, logged and described; transcripts were often shared with the other party. Every piece of post sent and received was filed in a colour-coded folder. In the end, Shockley's wife Emmy remained his sole companion. She was devoted, recalling in an interview later the one occasion in more than thirty years that Bill Shockley told her that he loved her.

At the age of seventy-seven, prostate cancer spread to Shockley's bones. He refused to call his children and in August 1989 he died, disgraced and alone except for Emmy. His daughter and two sons learned about their father's death from the newspapers. For the next ten years, Emmy Shockley kept the house exactly as it was on the day her husband died in 1989. At the time of his death, William Shockley had been named one of the '100 Most Influential People of the 20th Century' by *Time* magazine and had more than ninety patents to his name.

Business calls
ALEXANDER GRAHAM BELL

Born: March 3, 1847, Edinburgh, Scotland
Died: August 2, 1922, Baddeck, Nova Scotia

*The patent system added the fuel of interest to
the fire of genius.*
Abraham Lincoln

Don't let me be bound hand and soul to the Telephone.
Alexander Graham Bell

IF WHAT HAS BEEN CLAIMED ABOUT ALEXANDER GRAHAM BELL is true
– that his successful development of the telephone was based on a concept he stole
from Elisha Gray's confidential caveat filed earlier with the Patent Office – we are
seeing one of history's great myths crash and burn. How could that have happened?
Bell was a decent man, gentle and thoughtful with his students and profoundly caring
about the plight of the deaf, not a cunning and deceitful shyster.

In the small attic of a boarding house in Boston, Massachusetts, on March 10,
1876, Alexander Graham Bell spoke into the transmitter of his newly invented
telephone. The words his assistant Thomas Watson picked up on the receiver in the
next room, 'Come here, Watson, I need you,' were crude but clear.

In the wake of that call, Alec Bell demonstrated extraordinary vision in a letter to his
father. 'The day is coming,' Bell predicted, 'when telegraph wires will be laid on to houses,
just like water or gas, and friends will converse with each other without leaving home.'

Alexander Bell came by his passion for words honestly. Both his grandfather, Alexander, and his father, Melville, were experts in elocution and speech. His mother, Eliza, was almost completely deaf. With her ear trumpet pressed against the piano, she charted her son's progress as he learned to play, a skill that provided him with a great sensitivity to tone and pitch. Melville Bell developed a well-known system known as 'Visible Speech', a collection of symbols illustrating the correct shaping of the tongue, throat, lips and mouth for all sounds possible in the human voice, which was an important aid in teaching the deaf to speak.

In July 1870, having lost two of his three sons to tuberculosis, Melville Bell decided to move what remained of his family from England to the healthier climate of Canada. Within a year of arriving in Quebec, Alec had travelled to Boston, where he taught the deaf and lectured at Boston University. In his spare time, Alec Bell was also busy inventing.

In the throes of telegraph fever, America's entrepreneurial attention was focused on finding a way to send more than one message at a time across the wires, and Bell was among those working on new ideas. The race towards this lucrative patent was fierce and crowded with contenders, including Thomas Edison and the prolific inventor, Elisha Gray. Teaching by day and experimenting by night, Bell developed his theory for a 'harmonic' telegraph system that would send multiple messages at different pitches, a concept that attracted the attention of two wealthy Bostonians who were willing to back Bell's idea.

If there was foul play, it is likely that the seeds of that deceit started here – in the partnership between Bell, a highly strung, unsophisticated inventor; a wealthy investor, George Sanders; and the ruthless patent attorney and entrepreneur, Gardiner Hubbard. Obsessed with breaking the Western Union monopoly of the telegraph. Hubbard hungered to be the first to secure the multiple messaging system.

Under the partnership formed by these men, each was accorded a third of the profits arising from any successful patent. Hubbard and Sanders would bear the cost of an assistant for Bell – Thomas Watson – and other expenses, but no provision was made for Bell's personal needs, forcing him to work during the day as a teacher and run his experiments late into the night. The situation was further complicated when Alexander Bell fell deeply in love with Hubbard's daughter, Mabel, eleven years his junior and profoundly deaf. Adding to Hubbard's irritation, who had hoped for a more elegant suitor, the lovesick Bell was increasingly unable to concentrate on his inventing. From Bell's point of view, it was crucial to remain in the good graces of a future father-in-law whose overbearing ways were often oppressive. Once Mabel agreed to an engagement, Bell was faced with establishing himself financially in order to support a wife from well-to-do circumstances. When Gardiner Hubbard ran into financial difficulties of his own, the pressure on Bell to succeed increased many fold.

Pressed for cash, trying to meet a heavy teaching schedule and anxious over his beloved, Alexander Bell was further pressured by the awareness that others were moving forward quickly on improvements to the telegraph. Gardiner Hubbard advised Bell against applying for a caveat, the registering of a concept that had not yet been turned into a working prototype. Such a move would, Hubbard argued, give away too much to the competition. Instead, he urged Bell to work as swiftly as he could towards a patentable messaging system.

Letters written by Bell to his father indicate how excruciatingly anxious he felt in the face of his obligation to his backers, particularly as continued attempts to fashion the harmonic telegraph proved fruitless. Invention was no longer the tinkerer's dream, but a sharply competitive business with high stakes. Bell also experienced conflict over the direction of his experiments. His heart was really in his vision of an instrument that could transmit sound rather than Morse Code. Hubbard, focused on the telegraph, had no interest in a telephone, and ordered Bell to stop wasting his time on something so newfangled.

In 1875, Bell and Watson finally succeeded in sending three messages simultaneously across telegraph wires and Hubbard and Sanders despatched Bell to Washington to file a patent on his harmonic telegraph.

The mystery lies in a sequence of events that occurred eleven months later. On February 14, 1876, Gardiner Hubbard filed a patent on Bell's behalf for 'improvements to the telegraph', what was later known as the Telephone Patent. On the same day, Elisha Gray filed a caveat for his version of the telephone. On February 19, the Patent Office advised Bell that his patent would be suspended for three months because it 'interfered' with other applications. In other words, there was overlap with a caveat or a patent previously registered with the Patent Office, probably the caveat filed by Elisha Gray. Inexplicably, in spite of the three-month suspension, Bell was issued a patent on March 7, 1876. Three days later, Alexander Graham Bell spoke to Thomas Watson using a prototype that differed in important ways from the drawing Bell had submitted to the authorities only three weeks previously.

So how did Alexander Bell get past Elisha Gray's patent, and what was the source of the chages he made to his prototype? Ten years later, in 1886, Zenas Fisk Wilber, the Patent Examiner in charge of telegraphy when Bell filed in 1876, swore that he had shown Gray's confidential caveat to Bell just prior to Bell's breakthrough telephone call. Fisk also admitted that he had bypassed a formal investigation and arbitrarily declared Bell's patent the first to arrive in the office. Wilber was an admitted alcoholic and a man of highly questionable moral standards, but the explanation rings true. It is highly likely that Alexander Graham Bell was not the first to invent the telephone.

Less than three months after Bell's first successful transmission of sound over the wires, Gardiner Hubbard insisted that his future son-in-law demonstrate his new invention at the Philadelphia Centennial Exhibition. The members of the partnership, which now included Watson, were all badly in need of funds, and a successful demonstration at this all-important showcase for new inventions was critical. Everyone, including Edison and Elisha Gray, would be there.

The tension mounted as Bell persisted in his refusal to attend the Exhibition. Caught between her father and her fiancé, Mabel Hubbard eventually drove Bell to the train station, handed him a suitcase, and successfully implored him to board. Among those inspecting the exhibits was Dom Pedro, the Emperor of Brazil, who recognised Bell from an earlier meeting. Gingerly, the Emperor put the receiver to his ear and heard Bell, five hundred yards away, reciting from *Hamlet*. Before the day was over, Alexander Bell had won a gold medal for 'the greatest marvel hitherto achieved by the telegraph' as well as a gold medal for his display of Visible Speech.

Back at the hotel, Bell was informed that Elisha Gray had come to call. Meeting in Bell's room, they had a cordial conversation in the course of which Gray congratulated Bell on the 'glorious achievement of the vocal telegraph'.

Gardiner Greene Hubbard was not alone in his initial scepticism about Bell's invention. People were still adjusting to the wonder of the telegraph; what more did they need? Perhaps the telephone carried disease or, worse yet, involved some black magic. Perhaps everyone would be able to hear your private conversations. As miraculous as this new device was, interest in the telephone faded over the next months.

'It's an amazing invention,' said President Rutherford B. Hayes, referring to the telephone, 'but who would ever want to use one?' Prominent businessmen, including the powerful Thomas Orton of Western Union, did no better at envisioning the sea change. When the Bell Patent Association – Sanders, Hubbard, Bell and Thomas Watson – offered to sell the patent for the telephone to Western Union for $100,000, Orton dismissively turned down the offer, referring to the telephone as a 'mere toy'. Thomas Watson later claimed that within two years those patents were worth more than $25 million and Orton's decision went down as one of the worst business calls in history.

Everyone, including Mabel Hubbard, felt the urgency of turning the telephone into a lucrative commercial venture. Surprising everyone, the shy and awkward Alexander Bell, experienced from the days when he had helped his father demonstrate Visible Speech, took to the stage with an entertaining show for increasingly large and enthusiastic audiences. 'I am the invisible Tom Watson ' the audience would hear from the receiver. 'Everybody hears me! Nobody sees me!' Within a year, more than eight hundred telephones had been installed, a number that grew exponentially.

In July 1877, the same month in which thirty-year-old Alexander Graham Bell married Mabel Hubbard, he and his three partners formed the Bell Telephone Company. Alec's unusual wedding present to his new bride was his 30% share in the new entity, a remarkable gesture in a day when women were seldom involved in business. On their honeymoon trip to Europe, Alec Bell demonstrated his telephone for Queen Victoria.

In the more than six hundred lawsuits surrounding one of the world's most valuable patents, US Patent No. 174,465 for the 'speaking telegraph', Alexander Graham Bell was uniformly vindicated as the true inventor of the telephone. The first, and most challenging, of those suits pitted the Bell partnership against two of the industry's most powerful forces, Western Union and Thomas Edison.

Not long after their refusal to buy the patent, Western Union realised what they had passed up and sought to enter the market by a different route. Teaming up with Elisha Gray and twenty-seven-year-old Thomas Edison, who had developed a new and effective carbon transmitter, Western Union developed its own telephone system, which it pitched against that offered by Bell Telephone. In spite of the fact that the Bell partners were in dire financial straits, the telephone market had begun to stir and they chose to sue for infringement of their patent. The Bell Telephone Company, whose entire future was at stake, was pitted against a Goliath.

Summoned to testify, Alexander Graham Bell, still on honeymoon in Europe, railed about the pressures of being an inventor. His health declined. Again he refused to cooperate; again his wife Mabel persuaded him to fight. Arriving in Boston just in time and suffering from abscesses, Bell dictated his preliminary statement from his hospital bed at the Massachusetts General Hospital.

Rumours were rife. Newspapers reported that Elisha Gray was the true inventor of the telephone, that Bell had been less than honest in his dealings. Then, on April 7, 1879, the Bell Telephone Company lawyer rose and read a letter written by Elisha Gray to Alexander Graham Bell in March 1877.

'I do not claim even the credit of inventing it [the telephone]' Gray wrote, ' as I do not believe that the mere description of an idea that has never been reduced to practice ... should be dignified with the name invention.' That letter, which turned the tide, had been discovered in a small wastebasket in Alexander Graham Bell's attic workroom in Boston.

Inventing and teaching the deaf were the two things that Bell cared most about and he and Mabel now had sufficient money to allow him to indulge in both. He went on to invent a metal detector, with which he tried unsuccessfully to locate an assassin's bullet lodged in United States President James Garfield. When Bell's infant son died of breathing complications, Bell developed an early version of the artificial respirator, a metal jacket that fitted around the patient's body and applied and released pressure on the chest. This respirator was a precursor to the iron lung used later with polio patients.

In 1877, Thomas Edison invented the phonograph, but soon abandoned it as a useless invention. Four years later, Alexander Graham Bell and two other members of his Volta Laboratory team listened as their improved model played back its first words: 'I am a gramophone and my mother was a phonograph.' Edison was forced to purchase the Volta patent, including rights to the new wax cylinder, before he could put his phonograph on the market. Volta Laboratory, named after the lucrative and prestigious French prize awarded to Bell for the telephone, went on to patent the first phonograph records.

More patent suits, the death of two infant sons and of Mabel's two sisters, a growing distance from his family as Bell plunged into scientific work, and his chronic ill health gradually cast a shadow over the Bells. Life was too much with him. Then, in 1885, when he was thirty-seven and his wife twenty-six, Alec Bell made a decision. Life at their home in Washington was pleasant, but they both craved peace. Like Alexander's father and his mother Eliza, who had taken hold of life and moved from Britain to Canada, Alexander and Mabel packed their belongings and, with their two little girls, travelled by steamer and train to one of the world's largest saltwater lakes, Bras d'Or in Cape Breton. In the 1880s, Alec and Mabel Bell built a large estate in Baddeck, Nova Scotia, where they made their home happily for the next thirty-six years. Together, they had the wisdom to find a place where they could find peace and Alexander Graham Bell could pursue a life of pure invention.

Alexander Graham Bell's lifelong fascination with communication took another turn when he and his father-in-law, Gardiner Hubbard, invested in the eminent publication *Science*, started by Thomas Edison and abandoned by him when it proved to be unprofitable. Hubbard and Bell also established the National Geographic Society and started a new publication, *The National Geographic Magazine*. Gardiner Hubbard, who had only reluctantly agreed to his daughter Mabel's marriage to the socially insignificant Alexander Graham Bell, was to ultimately see all three of his daughters married to a member of the Bell family.

Teaching the deaf remained Alexander Graham Bell's first passion. One of those he helped was Helen Keller, who was both blind and deaf. Keller later described her meeting with the inventor as 'the door through which I should pass from darkness into light'.

In January 1915, Alexander Graham Bell and Thomas Watson reenacted their original conversation to mark the first telephone line to span the North American continent. "Mr Watson, come here, I need you."

"It would take a week for me to get to you this time," came the answer. Bell was in New York; Watson took the call in San Francisco.

On August 2, 1922, suffering from diabetes, Alexander Graham Bell died at his home in Nova Scotia. His body was buried in a simple pine coffin lined with kite fabric, in honour of his keen interest in the study of flight. On the rock that marks his grave were carved the words, 'Inventor – Teacher'. The telephone service in Canada and America was stopped for one minute in his honour.

Lightning line
SAMUEL FINLEY BREESE MORSE

Born: April 27, 1791, Charlestown, Massachusetts
Died: April 21, 1872, New York City

*Nineteenth-century wags were only half joking
when they declared that invention was the last resort
of all unsuccessful Americans.*

Carroll W. Pursell, Jr, *Those Inventive Americans*

THE DAY ITSELF, MAY 24, 1844, DIDN'T REALLY AMOUNT TO MUCH.
Samuel Morse, standing in the chamber of the Supreme Court in Washington, DC,
tapped out 'What hath God wrought?' along forty-four miles of poles and wire to
Alfred Vail, who was manning the receiver in Baltimore. A few dozen observers
exchanged messages before Morse signed off.

What did get people's attention a few days later was a blow-by-blow account
transmitted in Morse Code from the Democratic National Convention in Baltimore,
where dark horse candidate James K. Polk had been nominated for President. After a
dozen years of frustration and disappointment, Samuel Morse, then fifty-two years
old, enjoyed his first real triumph. Virtually redefining space and time, Morse had, it
was believed, 'chained the very lightning of heaven'.

Samuel Morse, the son of a Congregational minister, was an unlikely national
hero, a man ill-equipped for the fray that would surround an invention with as much
financial and commercial impact as the telegraph. Born in Charlestown,
Massachusetts, Morse went to Andover Academy as a boarder at the age of eight.

Communication with his mother and father consisted primarily of long letters of instruction on proper behaviour, education and religious development. Affection was doled out across the miles in direct proportion to good behaviour. There was respect but not warmth. 'The main business of life,' Jedediah wrote to his son, 'is to prepare for death.'

The first half of Morse's life was spent as a successful historical and portrait artist. Commissioned to paint a portrait of the Marquis de Lafayette, Morse was in Washington when he learned that his twenty-five-year-old wife had died, leaving three small children. The children, who were farmed out to friends and family, saw little of their father as they moved from pillar to post. Morse seldom answered their cries for attention, choosing instead to preach to them as his father had preached to him. 'I look to God to take care of you,' he wrote to his daughter, before going on to speak about his own grief.

As a student at Yale, Morse had been fascinated by the study of science. While outwardly engaged as a portrait painter, Morse worked quietly on a new means of communicating quickly over long distances using an electric battery and wires. Believing that his project was unique, it was a blow to read in 1837 that two English scientists, William Cooke and Charles Wheatstone, had patened a telegraph based on the principle of electromagnetism. Morse moved quickly to reveal his own, crude, invention. Receiving further reports of European advances in the telegraph, Morse shared the terrible fear that he would lose the race and, with it, all the financial and historical promise. Thus began a stressful, lifelong battle to establish himself as the inventor of the Magnetic Telegraph.

In October 1837, Morse filed a caveat with the Patent Office for the 'American Electro-Magnetic Telegraph', protecting his basic concept for a year. One aspect of Morse's idea was that of the relay, whereby a weak signal was able to open another strong circuit, propelling the message farther along the lines.

Before demonstrating his device for the first time, Morse instituted some important changes, including a technical concept first mooted by Joseph Henry, an eminent scientist and expert on matters electrical. This improvement based on Henry's work was the cornerstone of a lifelong battle over the value of Henry's contribution to the telegraph.

Alfred Vail, who joined the project at this crucial juncture, was another victim of

Morse's inability to be a team player. Vail provided his skills as a machinist as well as financial support in preparing the telegraph for its debut before Congress in 1844. In exchange, Vail was to receive one quarter of the revenue in the States as well as one half interest in the foreign rights. By all reports, Vail's mechanical contribution was significant, but when Vail and Morse demonstrated their system at the Franklin Institute in Philadelphia – at Vail's expense – only Morse's name appeared in print. When Morse met with President Van Buren in Washington, he went alone, without his 'assistant'.

While Vail worked his magic on the mechanism, Morse worked on the transmitting dictionary, in which a number was ascribed to every word used. Before long, he created the Morse Code, a faster system in which dots and dashes represented the letters themselves. The irony is that Morse's true invention, the Morse Code, was unpatentable. With patent law that allowed recognition of what was essentially software, Samuel Morse would have been the Bill Gates of his day.

Morse was not a good judge of men, he was not generous by nature, and he was not commercially astute. By the time he sent the first successful telegraphic message, Samuel Morse retained only nine-sixteenths of the patent rights in America and half of the overseas rights, having sold or traded the rest for financial backing and technical support in the twelve difficult years between the initial prototype and the successful transmission. Samuel Morse had no income and no reserves on which to call; in a word, he was broke. Partners included Francis O.J. Smith, Alfred Vail and Leonard Gale, Morse's scientific adviser.

It wasn't long before relationships between Morse and his partners became strained. Vail felt unappreciated and was frustrated by the delay in registering his two-sixteenths share of the patent with the authorities. In the event of Morse's death, Vail had no proof of ownership. The partners were equally disgruntled when Morse sold off patent rights, which everyone agreed were worth millions, for relatively small sums of money.

In 1843, Congress voted to provide the $30,000 Morse requested for a demonstration line between Baltimore and Washington. The penniless Morse was accorded an annual salary of $2,500. Morse was not so fortunate later, when the Postmaster General advised Congress to refuse Morse's offer to sell the telegraph to the government for $100,000. The telegraph, according to the government official, 'would never make money'.

Less than a year after the first transmission, Morse met journalist and political figure, Amos Kendall and, in one of his more enlightened decisions, turned over development of the telegraph project to this brilliant and honourable man. Within months, Kendall had formed or licensed companies to extend lines over huge networks, which paid dividends and royalties to Morse and his associates. Amos Kendall was one of the few men with whom Morse did not fall out, although there were some close calls.

The great success of his invention brought Samuel Morse the sort of recognition he had sought throughout his life, including an honorary LL.D. degree from his alma mater, Yale University. When a visitor demonstrated the wonders of the telegraph to the Sultan of Turkey by fixing a line between the main entrance and the royal harem, Morse was rewarded with Turkey's Order of Glory, a gold medallion studded with diamonds. Morse was the first American to be so honoured.

For the first time in his life, Morse faced the prospect of financial stability in the mid-1840s. In 1848, he met and married Sarah Griswold, a gentle deaf woman thirty years his junior and, in short order, they added four children to the three Morse already had. Home was Locust Grove, a hundred-acre farm on the Hudson River just outside Poughkeepsie, New York, on which he built a Tuscan-style villa.

However, any notion that Morse's troubles were behind him was short-lived. The financial potential presented by the extension of the telegraph and the powerful interests concerned with its success outran any plan for orderly development. Serious challenges to Morse and his patents arose and were thrashed out in court through long and costly litigation. In all, Samuel Morse struggled through seven intellectual property trials, each one requiring painstaking preparation. Adding insult to injury, Morse was vilified from every corner, not least by those who had helped him along the way and who now felt unacknowledged and unappreciated. Here, as with his children, Morse's austerity, religiosity, high-mindedness and inability to demonstrate affection often left those associated with him feeling undervalued.

Years of bitter courtroom battles culminated in the Great Telegraph Case argued before the Supreme Court and settled, for the most part, in Morse's favour in 1854, a decade after the first telegraphic demonstration. In identifying Morse as 'the first and original inventor of the Telegraph,' the Justices acknowledged that Morse had drawn on outside expertise, but agreed that 'the fact that Morse sought and obtained the necessary information and counsel from the best sources and acted upon it, neither impairs his rights as an inventor, nor detracts from his merits'.

Shortly thereafter, Morse was granted a seven-year extension of his patent on the grounds that much of his earnings, calculated at about $200,000, had been sacrificed to litigation. With the legal way cleared and thousands of miles of lines extending across America, Europe, Australia, India, Russia and China, it seemed that Morse could finally retreat to Locust Grove and enjoy the fruits of his labours.

Alfred Vail's comment that 'Professor Morse has confidence in everybody 'till they cheat him,' proved to be true once again as the issue of a transatlantic cable came under consideration. A wealthy American entrepreneur, Cyrus W. Field, formed the New York, Newfoundland and London Telegraph Company, of which Samuel Morse was a 10% shareholder and titular 'company electrician'.

Morse was flattered by his new colleagues and enthusiastic about a project he believed would enhance humanitarianism and end war between nations. Displaying the naivety that had frequently caused such anguish both to himself and to those who trusted in him, Morse granted the new transatlantic company free patent rights between New York and Maine, where the Atlantic cable was to begin. Now, as in the past, Morse turned a deaf ear to cautionary advice. 'Your true friends,' wrote Amos Kendall, Morse's longtime business adviser and loyal protector of the inventor's best interests, 'do not comprehend how it is that you give your time, your labor, and your fame to build up an interest deliberately and unscrupulously hostile to all their interests and to your own.'

It wasn't long before Morse's new business partners, led by Cyrus Field, betrayed him by purchasing the patent rights to a competitive telegraph system and proposed a new network in direct competition to Morse's own. On the edge of confrontation, Morse's desire to be part of the great adventure of the transatlantic cable was too compelling. Morse's background, which espoused genteel poverty over vulgar materialism, ill-prepared him to deal with or understand rough-and-tumble entrepreneurs like Field, referred to by Morse as 'mere men of trade'.

Morse did sail on the first attempt to lay the cable on the ocean floor in August 1857, but three hundred miles out the wire snapped and all was lost. With typical pomposity, Morse, who was not even on deck at the time, jumped into print, blaming a young British engineer for the disaster. Morse's title of 'honorary director' was withdrawn from the cable project and he was not included in further cable-laying attempts. Significantly, Samuel Morse was not present when Cyrus Field, aboard Brunel's *Great Eastern*, succeeded in laying the transatlantic cable in 1866. Morse received no mention in any of the newspaper reports or expansive congratulatory toasts.

The sale of his remaining patent rights in exchange for cash and stock in American Telegraph made Morse comfortably rich as he approached his seventieth year. With some of the proceeds he purchased a four-storey townhouse in New York City, which included a library, a gymnasium, a storey-high picture gallery and a conservatory. Later, as Western Union consolidated independent telegraph companies, Morse became wealthier still.

A chord of melancholy and depression runs through the Morse family, and Samuel was no exception. A devout and self-righteous Presbyterian, he was violently anti-Catholic and pro-slavery. As for Abraham Lincoln, he was 'coarse, vulgar, illiterate, inhuman, wicked and, above all, irreligious'.

Samuel Morse died leaving major debris in his wake. Friends, colleagues and family suffered from his self-centredness and lack of compassion. Taking to heart the pain of ridicule, betrayal and slights to his honour, Samuel Morse seemed incapable of comprehending that same pain in others. It says it all that when Alfred Vail's widow came to Morse begging for her husband's just due, she left the New York townhouse clutching only a photograph of Samuel Finley Breese Morse.

The portrait of Samuel Morse taken in 1863, his full white beard flowing over a chest laden with medals, shows him as the vainglorious man he was. Honours, accolades and titles meant more to him than almost anything else and Europe provided a rich source for the recognition and high society he craved. At seventy-five, he and his wife moved to an elegant apartment in Paris, at 10 avenue de Roi, where they employed a valet, a cook, a chambermaid and a seamstress.

Morse passed his heritage of benign neglect and pontification along to his seven children, the youngest of whom was four when he died. The consequences were sad. His son Charles, a ne'er-do-well, shared his father's gullibility. Morse's daughter Susan, at the age of sixty-six, disappeared on a passage to Havana, presumed to have jumped overboard to her death. Her son, an accomplished painter, committed suicide in his twenties. Another of Morse's grandsons hung himself from his children's swing. Arthur, a son from Morse's second marriage and a promising violinist, fell or jumped from a train in New Orleans and was killed instantly. Morse's son, Willie, was jailed for killing an Indian in Mexico; upon his release, he joined the Buffalo Bill Wild West Show.

Morse's name was a household word not only in America but worldwide, so much so that a letter addressed to 'Inventor of the Telegraph, New York City', reached his Gramercy Park brownstone, but the same seesaw between fame and ignominy followed Morse to his final days. A monument of the inventor joined those of Shakespeare and Schiller on the Central Park Mall, erected in his honour by Western Union. Dedication of the statue was followed by a sparkling banquet during which an operator tapped out Morse's chosen message, 'Glory to God in the highest, and on earth peace, good will to men', on the original telegraph machine. Morse tapped out his name to thunderous applause.

In the face of vehement accusations that Morse had usurped others' contributions to the telegraph, including Alfred Vail and Joseph Henry, the National Monument Association decided against erecting a similar statue in Washington. In February 1872, two months before Morse died, the chapter on the telegraph to be included in the popular book *Great Industries of the United States* arrived at Morse's bedside. He was devastated. In the account, Morse was reviled and dismissed as the true inventor of the telegraph, all his credits downgraded to delusion. In spite of his protests, the chapter was included in the book which sold ten thousand copies within a few months.

The praise and recognition Samuel Morse longed for was reserved for his obituaries. The outpourings of affection and praise – the national day of mourning, the memorial service at the House of Representatives attended by President Ulysses S. Grant, and the flags at half-mast – were lost to Morse forever.

An empty chair
SIR CHARLES WHEATSTONE

Born: February 6, 1802, Gloucester, England
Died: October 19, 1875, Paris, France

*Never did I feel such a tumultuous sensation
before as when, all alone in the still room,
I heard the needles click...*

Charles Wheatstone

AFTER APPROPRIATE TOASTS TO QUEEN VICTORIA, Prince Albert, and the President and People of the United States, William Fothergill Cooke rose and toasted the distinguished guests gathered in honour of Samuel Morse, who was visiting with his wife, Sarah. 'Mr Morse stands alone in America,' Cooke began, 'as the originator of the telegraph.' The simplicity of the system, '...conceived, worked out and perfected by him,' Cooke continued, 'has brought the Morse telegraph into use throughout Europe.'

Morse, in his turn, remarked that 'part of England' had been reluctant to acknowledge and honour the inventive American spirit, and that reluctance, he continued, had been 'a festering thorn in the hearts of some of the most cultivated in the land'.

There was an empty seat at this dinner in 1856. Charles Wheatstone, notable by his absence, had not received an invitation, the culmination of a bitter feud with his partner, William Fothergill Cooke.

Picture a young Charles Wheatstone, some fifty years before that dinner at the Royal Institution, fleeing from school, only to be captured on the streets of Windsor and returned to his home on Pall Mall. When next we see him, he is walking along the unpaved and dusty track that was Pall Mall in the 1820s, on his way to 436 The Strand, where he was apprenticed to his uncle, a maker and seller of musical instruments. In 1829, Wheatstone was awarded his first patent, for the concertina, an instrument belonging to the class of harmonicas and accordions.

Charles Wheatstone was a shy, sensitive boy whose first love was books of any and all kinds. At every opportunity, Charles haunted the bookstall near his home, where he spent any available pocket money. One book in particular caught his eye, a volume on Volta's electricity experiments written in French. He saved up his pennies and was finally able to buy not only the history, but a French dictionary to aid in translating the text.

Soon, Charles and his brother, William, retreated to the scullery with a home-made battery to try out Volta's findings for themselves. Numerous experiments on the transmission of sound ensued, during which Wheatstone conceived a plan for transmitting sound signals over long distances. Estimating that sound would travel two hundred miles a second through solid rods, he proposed a telegraph from London to Edinburgh and, at this early date, he called his arrangement a 'telephone'. He also devised a machine for augmenting feeble sounds, which he called a 'microphone'.

By 1834, when Wheatstone was appointed to the chair of experimental physics at King's College, London, he had a number of inventions under his wing, including the kaleidoscope and the concertina. Wheatstone's portable harmonium won a medal at the Great Exhibition of 1851, where he must surely have come across fellow inventors Charles Goodyear and Sam Colt.

Alone in the laboratory, Wheatstone was in his element. As a lecturer, he was tongue-tied and uncomfortable, often turning his back to the students to mumble instead at the blackboard. Called on to deliver the Friday Night Lecture at London's prestigious Royal Institution, Wheatstone suffered an attack of nerves and fled; Michael Faraday stepped in to fill the gap.

One of Wheatstone's experiments, the measurement of the velocity of electricity in a wire, was both beautiful and ingenious. In other experiments, Wheatstone calculated the speed of light and he also called attention to the value of thermo-electricity as a mode of generating a current by means of heat.

Then, in 1835, nine years before Samuel Morse's first demonstration of the telegraph in the United States, Charles Wheatstone developed an electric telegraph, proposing to lay an experimental line across the Thames. In 1837, he and William Fothergill Cooke devised the first practical telegraph, using an alphabetical system employing five needles: the needles would point to letters, which the operator would then note down. A patent was issued to them in that same year for the five-needle system. Their most important patent, describing an electric telegraph using only one magnetic needle, was patented in 1845. This telegraph was written up in the *Journal* of the Franklin Institute, where it caught Samuel Morse's attention.

When Morse sailed to Europe, Wheatstone invited him to King's College to view the telegraph. At the time, Morse was sending messages through a circuit of seventeen hundred feet; Wheatstone was transmitting across nineteen miles. Morse found Wheatstone likeable and intelligent, a 'most liberal and generous-hearted man, decidedly a man of uncommon genius,' but he described Wheatstone's telegraph as too complex.

At the height of summer 1837, Wheatstone sat by his instrument in a small, dingy room in Euston. By candlelight, he sent the first message to a group of gentlemen waiting eagerly by a receiver in Camden Town, one and a half miles away. William Cooke replied. 'As I spelled the words,' wrote Wheatstone later, 'I felt all the magnitude of the invention pronounced to be practicable beyond cavil or dispute.'

London and Northwest Railway officials, unimpressed by Wheatstone and Cooke's success, requested the removal of the 'new-fangled' telegraphic machine. Isambard Kingdom Brunel's Great Western Railway exercised greater wisdom and, in 1839, a line was raised on posts across the thirteen miles between Paddington and West Drayton station.

It took a violent death to draw attention to Wheatstone's technical feat. Early one morning, Sarah Hart was found murdered. The police suspected a Mr John Tawell as the culprit. When a man answering Tawell's description was seen entering a first-class carriage at Slough station, heading for London, the police telegraphed ahead to Paddington. When Tawell stepped onto the platform, a detective was waiting. Tawell was convicted and hanged. The publicity surrounding the case highlighted the practical use of the new telegraph technology.

The partnership between Cooke and Wheatstone was one of convenience. Cooke was an entrepreneur with his eye on a fortune; Wheatstone was an academic. When differences arose as to who had actually 'invented' the telegraph, ombudsmen Marc

Isambard Brunel and Professor Daniell of King's College issued a statement. 'It is to the united labours of two gentlemen,' they wrote, 'so well qualified for mutual assistance that we must attribute the rapid progress which this important invention has made during the five years that they have been associated.' In spite of the finding that both men should be accorded the recognition, relations between Cooke and Wheatstone disintegrated over the years. One recurring theme in their bitter correspondence was Cooke's complaint that Wheatstone's name appeared before his own on the English as well as the American patent.

Four months after Samuel Morse sent his first telegraph from Washington to Baltimore in 1844, Wheatstone, assisted by Mr J.D. Llewellyn, submerged a length of insulated wire in Swansea Bay and signalled through it to Mumbles Lighthouse. The Electric Telegraph Company was registered in 1845 and Wheatstone received £33,000.

Wheatstone, who found it so difficult to speak in public, was animated and talkative in private conversation, particularly when discussing the subjects closest to his heart. He set up a wire connection with the General Post Office from his home in Park Crescent, Regent's Park, and would often entertain visitors by sending and receiving messages from friends throughout Britain and Europe. One evening, he was observed talking excitably to Lord Palmerston on the capabilities of the telegraph. 'You don't say so,' exclaimed Palmerston, 'I must have you tell that to the Lord Chancellor.' Palmerston was heard to remark that 'a time was coming when a minister might be asked in Parliament if war had broken out in India and he would be able to reply, "Wait a minute, I'll telegraph the Governor General and let you know."'

One creative device after another flowed from Wheatstone's lab, earning him medals and recognition from the highest scientific quarters. Among those was the chronoscope, used for measuring minute intervals of time; a reflecting mirror stereoscope which presaged our 3-D films; an electro-magnetic clock; a solar clock, which was included in equipment used by Captain Nares' polar expedition; and the 'Wheatstone Bridge', the famous balance used to measure the electrical resistance of a conductor. In the course of perfecting his telegraphy system, Wheatstone noticed that wires charged with electricity often carried noises in a curious manner, a first step towards the telephone. Alexander Graham Bell, living with his grandfather in London, was invited to Wheatstone's laboratory, where he observed this phenomenon.

Wheatstone also wrote and decoded ciphers. His most famous, the Playfair Cipher, was named after his friend, the first Baron Playfair St Andrews, and was used by British forces in the Boer War and the First World War, and by the Australian Coastwatchers during the Second World War. He introduced paper tapes for the storage and transmission of telegraphic data, using two rows of holes to represent Morse's dots and dashes. By 1858, a Morse Paper Tape Transmitter could operate at a hundred words a minute.

For all his unquestioned brilliance and respectability, Wheatstone wasn't spared that which most inventors face at least once, a challenge to ownership of ideas. Wheatstone was charged with appropriating someone else's concept for the electro-magnetic clock and the electro-magnetic printing telegraph, but the inventor was able to prove satisfactorily that his work had preceded that of his assistant, William Fothergill Cooke.

The honours accorded Charles Wheatstone were capped by a knighthood, conferred on January 30, 1868. In autumn 1875, on a trip to Paris, Wheatstone caught a cold which produced inflammation of the lungs and he died on October 19, 1875 at seventy-three.

William Fothergill Cooke was knighted in 1869. Although he had earned a fortune in his lifetime, he died almost penniless on June 25, 1879. His relationship with Wheatstone, shattered over the controversy of invention rights, was never repaired, and the empty chair at Samuel Morse's London banquet stood as the symbol of friendship lost between two towering intellectuals.

Gentle giant
CHARLES PROTEUS STEINMETZ

Born: April 9, 1865, Breslau, Lower Silesia
Died: October 26, 1923, Schenectady, New York

If a young man goes at his work as a means of making money only, I am not interested in him. However, I am interested if he seems to do his work for work's sake, for the satisfaction he gets out of doing it. If I were to bequeath to every young man one virtue, I would give him the spirit of divine discontent, for without it, the world would stand still. The man hard to satisfy moves forward. The man satisfied with what he has done, moves backward.

Charles Proteus Steinmetz

IN THE LATE 1880s, A PENNILESS IMMIGRANT STOOD ALONE in New York City. Born in Breslau in 1865 and christened Carl August Rudolph Steinmetz, he was, like his father and his grandfather before him, deformed and dwarfish, with one leg twisted, a humped back and a large head on a thin body.

As a member of the Mathematical Society at the University of Breslau, Steinmetz had been given the nickname 'Proteus', after the kindly Greek sea god with the gift of prophecy. As a student, Steinmetz also joined the burgeoning political movement, attending secret meetings of the socialist group and distributing banned literature.

Warned that he was soon to be arrested for his political activities, Steinmetz was forced to flee Germany only months before earning his Doctorate. Early one morning in May 1888, he woke his father, explaining that he was going to visit a friend. Carrying a suit, a shirt or two, a few books and a copy of his thesis, Steinmetz crossed from Germany into Austria and onward to Zurich, where he spent the next year. He made friends quickly, compensating for his twisted body with an alert and pleasant manner, a strong handshake and a formidable intelligence. He would never see his father again.

In 1889, a Danish friend, Oscar Asmussen, urged Steinmetz to join him on the journey to America and offered to pay his way. Another friend provided Steinmetz with an introduction to the manufacturer and inventor, Rudolf Eickemeyer. On a Saturday afternoon in May 1889, *La Champagne* steamed into New York harbour and docked, but steerage passengers were not allowed to disembark on Castle Garden until Monday. Steinmetz stepped off the boat virtually penniless. 'You speak English?' the official asked him.

'A few,' came the hesitant answer.

'Do you have a job in America?' When Steinmetz didn't understand the question, the official repeated it angrily.

'Nein,' Steinmetz replied.

The official pointed to the forlorn figure, explaining to his superior that Steinmetz didn't speak English, had no money and no job and was sick and crippled. 'No,' the superior agreed, 'he can't come in.'

Steinmetz was ushered to the detention pen to wait for a boat returning to Europe. It was only through Asmussen's intervention and his pledge to guarantee his friend that Steinmetz finally passed through immigration and walked out into the sunshine of Battery Park, just four years after Nikola Tesla walked the same ground. Together, Asmussen and Steinmetz boarded the small ferry that crossed the East River under the shadow of the new Brooklyn Bridge.

Armed with a letter of introduction, Steinmetz first approached the head of the Edison machine works, but they made it clear that there were no openings for him there. When Steinmetz left, the chief engineer remarked to his associate that there was a regular epidemic of electricians coming to America.

The next day, twenty-four-year-old Charles Steinmetz travelled by train to Yonkers where he called on Rudolph Eickemeyer, inventor and owner of a business specialising in the production of hat-manufacturing machines. A clerk in the office remembered a man wearing plain, rough clothes and a cap, with nothing about his appearance to recommend him. Steinmetz asked to see Mr Eickemeyer and, quite remarkably given his appearance, was granted an interview. The two spoke in German for a couple of hours, sitting at Mr Eickemeyer's desk. All the latest news in electrical matters was discussed, including the most recent technical advances in transformers, storage batteries and magnetic apparatus. For Steinmetz, the discussion was largely theoretical, as he had never handled a direct current motor, and although he had written a technical paper on the theory of alternating current transformers, he had never actually seen one. The interview over, Eickemeyer took Steinmetz' particulars and offered to call him if an opening occurred.

A week later, Steinmetz again presented himself at the Eickemeyer establishment. His persistence paid off, and he was told to report to work as a draftsman the following Monday, June 10, 1889. His pay was $12 a week and his first task was the drafting of street car motor number 3. Wishing to be more American, he changed his name from Carl to Charles, and, noticing that most people had a middle initial, he called on his nickname from university days and added the 'P' for Proteus. Charles Proteus Steinmetz it was.

At the time Steinmetz arrived in Yonkers, Rudolph Eickemeyer was a man of means, living in a large brick house and employing a number of servants. Tall and straight with a flowing beard, Eickemeyer was highly dignified; he was also a born manager who encouraged his workers and nurtured any trace of originality in their thinking. Charles Steinmetz was a regular guest at Eickemeyer's Sunday evening dinners, where the table stretched from wall to wall of the large house and men of all disciplines gathered to talk. Steinmetz, admired and liked by all, was so short that his head just rose above the table top. To avoid asking others to pass him a dish, he would balance himself on the rungs of his chair, lunge forward, fork high in the air, spear something and then drop back into his chair.

Rudolf Eickemeyer recognised genius and, relieving his new employee of pedestrian tasks, he set Steinmetz up in a small laboratory. Steinmetz' tender nature was exemplified one day when a visiting engineer climbed the three flights of stairs to find a cold room and the scientist working in his overcoat and hat. "Mr Steinmetz," he finally asked, "why don't you build a fire in the stove?" "Oh, that," was the reply. "A mouse has had her babies in there and they are not yet old enough to move."

It was in that unimposing laboratory that Charles Steinmetz pioneered the critical law of hysteresis, a fundamental principle of alternating magnetism that enabled engineers to determine power loss under specific conditions. Tests in his laboratory produced a comprehensive table of constants for all magnetic materials known at the time. This innovative theory, first published in 1891, just three years after his arrival in the United States, was crucial to engineers as they worked to maximise the efficiency of alternating current. Charles Steinmetz had provided a mathematical tool for reducing alternating current theory to a series of practical calculations.

On January 19, 1892, Steinmetz presented his paper on hysteresis to the prestigious American Institute of Electrical Engineers. Attendance was light and there was nothing to indicate the importance of the occasion. The speaker was odd in appearance, dressed in an ill-fitting, unpressed suit and wearing overshoes into which he tucked the bottom of his trousers. He delivered his talk in a high, singsong voice. The complex mathematics were one thing; communicating them was another. When Steinmetz presented his theory to the International Electrical Congress in 1893, a large part of his audience walked out. The mathematical formulas that he put forward to achieve production and distribution of alternating current were so subtle and complex that few could understand them.

As a pioneer in mathematical theory, Steinmetz was to see nearly four years pass before his ideas were published, but he was a patient and unassuming teacher as well as a mathematical genius. Not least, he carried the entire table of logarithms up to a thousand in his head and knew by heart the seven-point extensions of this table. He was able to calculate intricate mathematical problems in seconds without paper or pencil. Some began to refer to him as a wizard but, unlike Edison, Steinmetz did not like the implication that he was above his fellow man.

At about the same time that Steinmetz presented his paper on hysteresis, the Edison General Electric Company of New York and the Thomson-Houston Electric Company of Lynn, Massachusetts merged to form the General Electric Company. E.W. Rice, the future president of General Electric, visited Steinmetz in Eickemeyer's laboratory. The man who greeted him, sitting cross-legged on a laboratory work table, was far from imposing. The gnome-like Steinmetz had a large head atop a frail body and his long hair fell to his shoulders. He wore an old cardigan and his trademark cigar hung from his mouth. Like others, though, Rice found that the minute Steinmetz spoke, all reservations about his appearance fell away. Like Eickemeyer before him, Rice quickly saw beyond the eccentric figure to a man of brilliance and kindliness. Here, explained Rice later, was a great man who spoke with the authority of accurate and profound knowledge, a man who brought order out of chaos in the matter of alternating current calculations. Rice wasted no time in recommending that General Electric offer Steinmetz a job. He turned them down. He could not, he told them, leave the man who had been so good to him. So important was Steinmetz to their plans that General Electric bought the firm of Eickemeyer and Osterheld in 1893 in order to secure its prize employee.

Fortunately, General Electric had the wisdom to let Steinmetz do what he did best. They provided him with freedom of research and development, the tools he needed and a laboratory in which he could work. They also accepted his preference for working outside the General Electric plant, allowing him full latitude to come and go as he pleased. Far more than money, Steinmetz valued the freedom to work when and where he wanted and to focus on a project of choice. His tools were an old Nabisco tin filled with sharpened pencils and a supply of plain paper.

Steinmetz was not an inventor in the mould of Edison; his forté was the theory that facilitated practical invention. Projects to which Steinmetz turned his inventive mind included a method of high-tension transmission, dynamos, the magnetic arc lamp and a motor for the Otis elevator. One of his most important discoveries was a means of protecting power lines, transformers and generators from the damage inflicted by lightning. One winter's day in 1922, a series of violent explosions rocked the lab and Steinmetz emerged, smiling; among the attendees at the demonstration of the lightning arrester was Thomas Edison. At his death, Charles Steinmetz held nearly a hundred and ninety-five patents, forty of which pertained to alternating current and its distribution.

Home in Schenectady, New York, where General Electric was based, comprised rented accommodation shared with two other bachelors. It was here that Steinmetz began to build his collection of pets, including two black crows, John and Mary, with which he had lengthy conversations. When the crows met an untimely end, Steinmetz had them stuffed and installed on the bookshelf. In addition to owls, eagles and cranes, the menagerie included a monkey named Jenny and a pet Gila monster that roamed the laboratory, dying eventually, as Steinmetz claimed, from lack of ambition.

One afternoon, the Schenectady police received an excited call. 'No madam,' the officer exclaimed, 'we do not have alligators in the Erie Canal.' On closer inspection, there were, indeed, two alligators swimming in the canal – and they belonged to Charles Steinmetz. As nets were lowered, Steinmetz appeared on his bicycle. The crowd, still not used to his odd appearance and suspicious about his activities, drew back.

'Ach,' he exclaimed, looking over the bridge. 'They wanted a swim.'

When one of the reptilian wanderers was brought out thrashing, Steinmetz cradled it in his arms, chatted softly and rubbed its belly. Within minutes, the alligator was calm.

Steinmetz also collected exotic plants, including hundreds of varieties of cacti, many of which in their twisted and strange forms resembled himself. Every spare moment from early March until late autumn, Steinmetz spent at his rustic camp on Viele's Creek, a branch of the Mohawk River. Even before the ice left the creek, he was back in his old red bathing suit. Visitors to the camp, no matter how esteemed, were apt to find Steinmetz drifting on the river in his battered twelve-foot canoe, papers spread out in front of him on boards. In parting, everyone received his characteristic 'so long'.

For a while, the townspeople looked askance at Steinmetz and his curious habits, but it wasn't long before the people at General Electric. formed true affection for their colleague. He was, they discovered, never too busy or too proud to stop and help someone with a problem or to share his knowledge and expertise. Eventually, the local people, impressed by reports and articles about their resident genius, accepted Charles Steinmetz wholeheartedly, referring to him as 'our Steinmetz'.

In 1900, Charles Steinmetz befriended a young lab assistant, Joseph LeRoy Hayden, who also worked at General Electric. Hayden soon occupied a few rooms in his friend's new house. The close relationship that developed relieved Steinmetz of his deep-seated loneliness, so Hayden's announcement that he intended to marry came as a blow. Much of Steinmetz' enthusiasm for his new, three-storey house and laboratory waned as he found himself once again alone. Hayden married in May 1903 and, after a week's honeymoon, Hayden and his bride Corinne returned to their small apartment. Later that day, Charles Steinmetz knocked on their door and he and his assistant sat down to smoke their cigars and discuss work.

Before long, Steinmetz invited the couple to move in with him. Mrs Hayden had reservations and stated them, but the plan went ahead and Charles Proteus Steinmetz had a family for the first time in his life. Corinne Hayden imposed rules, not least about the sharing of expenses, but the man who lost himself in work, forgot to take baths and missed appointments with important people was hard to manage. Over time, Corinne came to appreciate their mentor as a cheerful and loving person, one who accepted people as they were. He referred to her by a pet name, 'Mousie'.

One day, Charles Steinmetz marched into the kitchen and urged Corinne outside. There stood an electric car, complete with tasselled curtains and a cut-glass flower vase. "Who is the car for?" she asked.

"I thought I needed one," he replied.

"But you don't drive," she pointed out.

"Oh," he said, waving his hand, "someone will."

This new and very happy group, which soon included the Hayden's three children, endured until the end of Steinmetz' life some twenty years later, providing the last link — a family — for this man who was never to marry. The day that the Haydens' first son was born Steinmetz asked Corinne if she approved of his adopting her husband formally as his son. 'I approve, Grandpa,' she said, extending her hand.

One of the famous guests who visited the Steinmetz house on Wendell Street was Henry Ford, who came to consult him about a problem they were having with the headlamps on the new Ford motor car. In the middle of their discussion, young Midge Hayden burst in, requesting a bedtime story. Steinmetz rose and excused himself for half an hour. Ford, not used to being dismissed, was infuriated as he paced the floor to pass the time, but by the time Steinmetz returned, he had found the solution to Ford's problem.

At eight o'clock on the morning of October 26, 1923, Charles Proteus Steinmetz died at the age of fifty-eight of a heart attack at his home in Schenectady. Thousands of people gathered at his grave and the pall bearers included Owen Young, chairman of the board of General Electric, and Gerard Swope, the company's president. To everyone's surprise, Steinmetz left no great estate. Charles Proteus Steinmetz, a socialist and one of the world's greatest mathematicians and electrical engineers, believed throughout his life that no man should accumulate more than he required for his daily needs. Work was its own reward and what money he didn't require for his own needs, he was happy to give to others.

Mrs Hayden, part of Steinmetz' adopted family for twenty-one years, described him thus, 'a man with a brilliant intellect, a big and loving heart and a soul as white as a child's'.

Bibliography

Adams, Russell B. Jr. *King C. Gillette: The Man and His Wonderful Shaving Device*. Boston and Toronto, Little Brown and Company, 1978.

Baldwin, Neil. *Edison: Inventing the Century*. New York, Hyperion, 1995.

Brandon, Ruth. *Singer and the Sewing Machine: A Capitalist Romance*. Philadelphia and New York, J.B. Lippincott Company, 1977.

Bridgman, Roger. *Inventions and Discoveries*. London Dorling Kindersley Ltd 2002.

Brown, G.I. *The Big Bang: A History of Explosives*. Gloucestershire, Sutton Publishing Ltd, 2005.

Buchanan, Angus. *Brunel: The Life and Times of Isambard Kingdom Brunel*. London and New York, Hambledon and London, 2002.

Burlingame, Roger. *Henry Ford: A Great Life in Brief*. New York, Alfred A Knopf, 1955.

Carson, Mary Kay. *Alexander Graham Bell: Giving Voice to the World*. New York and London, Sterling Biographies, 2007.

Carter, Richard. *Breakthrough: The Saga of Jonas Salk*. New York, Trident Press, 1966.

Cheney, Margaret. *Tesla: Man Out of Time*. New York, Dorset Press, 1981.

Conot, Robert. *A Streak of Luck*. New York, Seaview Books, 1979.

Dessauer, John H. *My Years with Xerox: The Billions Nobody Wanted*. New York, Manor Books, Inc., 1974.

Dyson, James. *James Dyson's History of Great Inventions*. London, Constable and Robinson Ltd, 2001.

Edwards, Florence. *Cats Eyes*. Oxford, Blackwell, 1972.

Evans, Harold. *They Made America*. New York, Little Brown & Company, 2004.

Fant, Kenne. *Alfred Nobel: A Biography*. New York, Arcade Publishing, 1991.

Feldman, Burton. *The Nobel Prize: A History of Genius, Controversy and Prestige*. New York, Arcade Publishing, 2000.

Field, Colonel C. *The Story of the Submarine*. Philadelphia, J.B. Lippincott Company, 1908.

Gibbs-Smith, C.H. (ed.) *The Great Exhibition of 1851: A Commemorative Album*. London, His Majesty's Stationery Office, 1950.

Gillette, King C. *The Human Drift* (1894) with an introduction by Kenneth M. Roemer. New York, Scholars' Facsimilies & Reprints, 1976.

Grant, Ellsworth S. *The Colt Legacy: The Colt Armory in Hartford, 1866–1980*. Providence, Rhode Island, The Mowbray Company, 1982.

Green, Constance. *Eli Whitney and the Birth of American Technology*. Boston, Little, Brown, 1956.

Grosser, Morton. *Diesel: The Man and the Engine*. New York, Athenaeum, 1978.

Groundwater, Jennifer. *Alexander Graham Bell: The Spirit of Invention*. Alberta, Canada, Altitude Publishing, 2005.

Gunther, Ralph. *The Magic Zone: Sketches of the Nobel Laureates*. Potomac, Maryland, Scripta Humanistica, 2003.

Halasz, Nicholas. *Nobel: A Biography of Alfred Nobel*. New York, The Orion Press, 1959.

Hallock, Robert Lay. *Inventing for Fun & Profit: A Manual on How to Develop, Protect, and Sell a Patentable Idea at Minimum Expense, and with Maximum Profit and Satisfaction*. New York, Harmony Books, 1978.

Hammond, John Winthrop. *Charles Proteus Steinmetz: A Biography*. New York and London, The Century Co., 1924.

Harrison, Ian. *The Book of Inventions*. Washington, DC, National Geographic, 2004.

Hubert, Philip G. Jr. *Men of Achievement*. New York, Charles Scribner's & Sons, 1896.

Iles, George. *Leading American Inventors*. New York, Henry Holt and Company, 1912.

Israel, Paul. *Edison: A Life of Invention*. New York, Wiley & Sons, Inc. 1998.

Jeans, W.T. *Lives of the Electricians: Professors Tyndall, Wheatstone and Morse*. London, Whittaker & Co., 1887.

Kirby, Richard Shelton (ed.). *Inventors and Engineers of Old New Haven: A Series of Six Lectures in 1938 under the auspices of the School of Engineering, Yale University*. New Haven, Connecticut, Colony Historical Society, 1939.

Larsen, Egon. *Ideas and Invention*. London, Spring Books, 1960.

Leonard, Jonathan Norton. *Loki: The Life of Charles Proteus Steinmetz*. Garden City, New York, Doubleday, Doran & Company, Inc., 1929.

Levinovitz, Agneta Wallin and Ringertz, Nils (eds) *The Nobel Prize: The First 100 Years.* London, Imperial College Press and World Scientific Publishing Co. Pte. Ltd, 2001.

McKibben, Gordon. *Cutting Edge: Gillette's Journey to Global Leadership.* Boston, Massachusetts, Harvard Business School Press, 1998.

Millard, Andre. *Edison and the Business of Invention.* Baltimore and London, The Johns Hopkins University Press, 1990.

Miller, Floyd. *Electrical Genius of Liberty Hall: The Man Who Tamed Lighting.* New York, McGraw-Hill, 1962.

Morris, William Knowles. *John P. Holland 1841–1914.* Columbia, South Carolina, University of South Carolina Press, 1966.

National Geographic Special Publications Division, *Those Inventive Americans.* 1971

Nitske, Robert and Wilson, Charles Morrow. *Rudolf Diesel: Pioneer of the Age of Power.* Norman, Oklahoma, University of Oklahoma Press, 1965.

Noble, Celia Brunel. *The Brunels: Father and Son.* London, Cobden-Sanderson, 1938.

Nuland, Sherwin B. *The Doctors' Plague: Germs, Childbed Fever, and the Strange Story of Ignaz Semmelweis.* New York and London, W. W. Norton & Company, 2003.

Nuland, Sherwin B., MD, FACS. *The Origins of Anesthesia.* Birmingham, Alabama, The Classics of Medicine Library, 1983.

O'Neill, John J. *Prodigal Genius: The Life of Nikola Tesla.* New York, Ives Washburn, Inc., 1944.

Oakley, Violet. *Samuel F.B. Morse: A Dramatic Outline of the Life of the Father of Telegraphy & The Founder of the National Academy of Design.* Chestnut Hill, Pennsylvania, The Eldon Press, 1939.

Okazaki, Manami. 'Retail Therapy: Fishing for Pearls', *Jetstar Airways Inflight Magazine.* Singapore, Ink Publishers, 2008.

Oshinsky, David M. *Polio: An American Story.* Oxford and New York, Oxford University Press, 2005.

Owen, David. *Copies in Seconds: Chester Carlson and the Birth of the Xerox Machine.* New York, Simon & Schuster, 2004.

Peirce, Rev. Bradford K. *Trials of an Inventor: Life and Discovery of Charles Goodyear.* New York, Carlton & Porter, 1866.

Pugsley, Sir Alfred (ed.) *The Works of Isambard Kingdom Brunel.* London, Institute of Civil Engineers and the University of Bristol, 1976.

Rawlence, Christopher. *The Missing Reel.* New York, Atheneum, 1990.

Reis, Ronald A. *Jonas Salk Microbiologist.* New York, part of the series Ferguson Career Biographies, 2006.

Robertson, Charles J. *Temple of Invention: History of a National Landmark.* London, Scala Publishers Limited, 2006.

Salk, Jonas. *Anatomy of Reality: Merging of Intuition and Reason.* New York, part of the Convergence Series edited by Ruth Nanda Anshen, Columbia University Press, 1983.

Shurkin, Joel N. *Broken Genius: The Rise and Fall of William Shockley, Creator of the Electronic Age.* New York, London, Melbourne and Hong Kong, MacMillan, 2006.

Silverman, Kenneth. *Lightning Man: The Accursed Life of Samuel F.B. Morse.* New York, Alfred A. Knopf, 2003.

Slack, Charles. *Noble Obsession: Charles Goodyear, Thomas Hancock, and the Race to Unlock the Greatest Industrial Secret of the Nineteenth Century.* New York, Hyperion, 2002.

Sohlman, Ragnar and Schück, Henrik. *Nobel: Dynamite and Peace.* Translated by Brian and Beatrix Lunn. New York, Cosmopolitan Book Corporation, 1929.

Steinberg, Neil. *Complete & Utter Failure: A Celebration of Also-Rans, Runners-Up, Never-Weres and Total Flops.* New York, Doubleday, 1994.

Tanaka, Junzo. 'Kansai Topics', *Kippo News.* Vol. 15, No. 572, July 23, 2008.

Tedlow, Richard S. *Giants of Enterprise: Seven Business Innovators and the Empires They Built.* New York, HarperBusiness, 2001.

Vare, Ethlie Ann and Ptacek, Greg. *Mothers of Invention.* New York, William Morrow and Company, Inc., 1988.

Wachhorst, Wyn. *Thomas Alva Edison: An American Myth.* Cambridge, Massachusetts and London, The MIT Press, 1981.

Waite, Helen L. *Make a Joyful Sound.* Philadelphia, Macrae Smith Company, 1961.

Whiting, Jim. *Auguste & Louis Lumière and the Rise of Motion Pictures.* Hockessin, Delaware, Mitchell Lane Publishers, 2006.

Williams, Trevor I. *A History of Invention.* London, Boston, New York, Little, Brown & Company, 1987.

Wolf, Ralph F. *India Rubber Man: The Story of Charles Goodyear.* Caldwell, Ohio, The Caxton Printers, Ltd, 1939.

Woodside, Martin. *Thomas Edison: The Man Who Lit Up the World.* New York, Sterling Publishing, 2007.

Photo credits

Page 2: 10/19/27: King C. Gillette as he returned from a trip abroad on the S.S. Homeric. © Bettmann/CORBIS, All Rights Reserved.

Page 8, 11: Courtesy of Xerox Corporation.

Page 19: Isaac Merritt Singer's first sewing machine, from *Genius Rewarded or the Story of the Sewing Machine*, New York, 1880. Image Select / Art Resource, NY.

Page 24: Kokichi Mikimoto, the sole inventor of cultured pearl system, standing near lagoon at his factory. (Photo by Alfred Eisenstaedt//Time Life Pictures/Getty Images). © Time Life Pictures.

Page 28: UPPA/Photoshot.

Page 32: Alfred Nobel, 1853 © The Nobel Foundation.

Page 40, 43: Kean Collection/Hulton Archive/Getty Images © 2004 Getty Images.

Page 46: From the collections of the Texas/Dallas History division. Dallas Public Library.

Page 50: Image no.978 *Sir Marc Isambard Brunel* James Northcote oil on canvas, 1812–1813, dated 1813. 49 1/4 in. x 39 in. (1251 mm x 991 mm) Given by the sitter's grandson, Henry Marc Brunel, 1895 © National Portrait Gallery, London.

Page 56: Image no. P112. *Isambard Kingdom Brunel* Robert Howlett albumen print, 1857 11 1/4 in. x 8 7/8 in. Given by Mr and Mrs A J W Vaughan, 1972 © National Portrait Gallery, London.

Page 66: John Holland climbing up the hatch of his invention; the submarine USS Holland. World Telegram Photo Courtesy of the U.S. Naval Historical Center.

Page 72: Fotografie von Franz Werner, Munchen 1883. Foto Deutsches Museum.

Page 80: Courtesy of George Eastman House, International Museum of Photography and Film.

Page 84: © John Hedgecoe/Topham/The Image Works.

Page 88: Engraving by W. B Jackman, D. Appleton & Company, NY. Library of Congress Prints and Photographs Division Washington, D.C.

Page 96: Library of Congress Prints and Photographs Division Washington, D.C.

Page 102: Courtesy of Massachusetts General Hospital Archives and Special Collections. Photo of Mezzotint.

Page 112: Jonas Salk, MD; Virus Research Lab, University of Pittsburgh; 1954. Courtesy of March of Dimes Foundation.

Page 122: Ventriloquist Paul Winchell with his dummy Jerry Mahoney. (Photo by Ray Fisher//Time Life Pictures/Getty Images) © Ray Fisher.

Page 125: © Greenberg/Ventura County Star.

Page 128: Albert C. Barnes c. 1900. The Barnes Foundation.

Page 134: Culver Pictures.

Page 138: Library of Congress Prints and Photographs Division Washington, D.C. H Bros., New York.

Page 150: Library of Congress Prints and Photographs Division Washington, D.C.

Page 158: Le Prince Getty Images. Photo by Hulton Archive/Getty Images.

Page 16: National Portrait Gallery, Smithsonian Institution / Art Resource, NY.

Page 164: © Time Life Pictures. Time & Life Pictures/Getty Images.

Page 170: American Institute of Physics Emilio Segre Visual Archives, Physics Today Collection.

Page 176: Lobrichon, Timoléon, artist. 'Alexander Graham Bell, three-quarter length portrait, standing, facing left.' Between 1882 and 1960. Gilbert H. Grosvenor Collection of Alexander Graham Bell photographs, Prints and Photographs Division, Library of Congress. Library of Congress Prints and Photographs Division Washington, D.C.

Page 184: Matthew Brady Studio. Library of Congress Prints and Photographs Division Washington, D.C.

Page 192: Hulton Archive/Getty Images.

Page 198: Library of Congress Prints and Photographs Division Washington, D.C.

Index